MATH/STAT.

NONPARAMETRIC
PROBABILITY DENSITY
ESTIMATION

Johns Hopkins Series
in the Mathematical Sciences
NUMBER

1

Nonparametric Probability Density Estimation

Richard A. Tapia
James R. Thompson

The Johns Hopkins University Press · Baltimore and London

7214-3514

MATH-STAT.

Repl. QA 273.6
T37
math

Copyright © 1978 by The Johns Hopkins University Press

All rights reserved. No part of this book may be reproduced or transmitted in any form or by any means, electronic or mechanical, including photocopying, recording, xerography, or any information storage and retrieval system, without permission in writing from the publisher. Manufactured in the United States of America.

The Johns Hopkins University Press, Baltimore, Maryland 21218
The Johns Hopkins Press Ltd., London

Library of Congress Catalog Number 77–17249
ISBN 0–8018–2031–6
Library of Congress Cataloging in Publication data will be found
on the last printed page of this book.

QA273
.6
T37
MATH

To my mother, Magda Tapia,
and my father Amado B. Tapia
—RICHARD A. TAPIA

To my mother, Mary H. Thompson,
and the memory of my father, Clarence A. Thompson
—JAMES R. THOMPSON

Contents

PREFACE   ix

ACKNOWLEDGMENTS   xi

CHAPTER 1. *Historical Background*   1

1.1. Moses as a Statistician   1
1.2. The Haberdasher and the "Histogram"   2
1.3. The Pearson Distributions   5
1.4. Density Estimation by Classical Maximum Likelihood   13

CHAPTER 2. *Some Approaches to Nonparametric Density Estimation*   24

2.1. The Normal Distribution as Universal Density   24
2.2. The Johnson Family of Distributions   30
2.3. The Symmetric Stable Distributions   33
2.4. Series Estimators   36
2.5. Kernel Estimators   44

CHAPTER 3. *Maximum Likelihood Density Estimation*   92

3.1. Maximum Likelihood Estimators   92
3.2. The Histogram as a Maximum Likelihood Estimator   95
3.3. The Infinite Dimensional Case   99

CHAPTER 4. *Maximum Penalized Likelihood Density Estimation*   102

4.1. Maximum Penalized Likelihood Estimators   102
4.2. The de Montricher–Tapia–Thompson Estimator   106
4.3. The First Estimator of Good and Gaskins   108
4.4. The Second Estimator of Good and Gaskins   114

CHAPTER 5. *Discrete Maximum Penalized Likelihood Estimation*   121

5.1. Discrete Maximum Penalized Likelihood Estimators   121
5.2. Consistency Properties of the DMPLE   124
5.3. Numerical Implementation and Monte Carlo Simulation   129

APPENDIX I. *An Introduction to Mathematical Optimization Theory*   146

I.1. Hilbert Space   146
I.2. Reproducing Kernel Hilbert Spaces   152
I.3. Convex Functionals and Differential Characterizations   154
I.4. Existence and Uniqueness of Solutions for Optimization Problems
    in Hilbert Space   160
I.5. Lagrange Multiplier Necessity Conditions   163

APPENDIX II. *Numerical Solution of Constrained
Optimization Problems*   166

II.1. The Diagonalized Multiplier Method   166
II.2. Optimization Problems with Nonnegativity Constraints   170

INDEX   173

Preface

This study was originally motivated by some difficulties encountered in the analysis of remote sensing data by standard maximum likelihood techniques. We quickly perceived that the underlying nonrobustness of classical procedures in the analysis was due to the presence of asymmetrical and multimodal densities rather than to tail pathologies. Consequently, neither the standard nonparametric techniques nor those of the Princeton Robustness Study were applicable.

In Section 21.1 of Fisher's *Design of Experiments*, the author questioned the utility of the philosophy of "nonparametric" testing. He had a point. Whenever the assumptions of standard nonparametric procedures (e.g., density symmetry, unimodality) are valid, it is usually not too optimistic to assume that the Central Limit Theorem is percolating benevolently. In our opinion, too many workers in robustness have been laboring under the assumption that if one only protects himself against pathology in the tails, then all will be well.

This study considers density estimation techniques in which very little is assumed. We have been most concerned with maximum penalized likelihood estimation methods. Consequently, the material is organized in a more or less historical fashion, pointing toward MPLE. We have considered several other techniques *en passant*, particularly, kernel density estimation.

As a reference text, our study is oriented toward the applied worker who is confronted with nonstandard densities (which is common) and realizes the fact (which is not). We have attempted to keep the book as self-contained as possible, with an appendix devoted to the necessary optimization-theoretic machinery. Software packages are already available (e.g., that of IMSL) which bring penalized likelihood within the realm of practical usage.

One semester graduate statistics courses, using the book as text, have been offered at Princeton and Rice universities. Portions of the material have been taught at Stanford University. Students had various backgrounds: computer science, engineering, numerical analysis, operations research, and statistics.

Houston and Princeton

Acknowledgments

The present study is an explication of results presented in ten invited lectures at The Johns Hopkins University in June 1976. We are indebted to Roger Horn, chairman of the Department of Mathematical Sciences at Johns Hopkins, for proposing the series of lectures and this book. To David Pyne and Eli Naddor we extend our thanks for organizing and implementing the logistics of the symposium.

Much of the work covered here follows from the doctoral dissertations of two of our former Rice University graduate students: Gilbert de Montricher and David Scott. All the figures in this book are the work of David Scott. A portion of the material in Section 2.5 has appeared in *Nonlinear Analysis* 1(1977): 339–72, in which Scott was coauthor. A part of this work is related to the doctoral dissertation of former Rice University graduate student John Bennett. We would like to thank all Rice graduate students and former graduate students with whom we have discussed this material—most notably, Richard Byrd, Rodney Grisham, and Lubomyr Zyla. Also, at Princeton, Roberta Guarino, Yoav Benjamini, Donna Mohr, and Allen Wilks pointed out a number of discrepancies in the first draft of the manuscript.

We have benefited greatly from discussions with our colleagues R. J. P. de Figueiredo, Paul Pfeiffer, Salomon Bochner, William Veech, John Tukey, I. J. Good, Jean-Pierre Carmichael, Emmanuel Parzen, Murray Rosenblatt, Chris Tsokos, and Grace Wahba.

The Office of Naval Research has encouraged our work since 1972 under contract NR–042–283. We have also received support from the Air Force under AFOSR 76–2711, from the National Science Foundation under ENG 74–17955, and from ERDA under E–(40–1)–5046. The NASA remote sensing data in Figures 5.8 through 5.12 was supplied by David van Rooy of Rice University and Ken Baker of NASA. Jean-Pierre Carmichael of the State University of New York, Buffalo, supplied the snowfall data in Figure 5.7. To Idalia Cuellar and Velva Power we extend our sincere thanks for their patient and competent typing of the manuscript.

NONPARAMETRIC

PROBABILITY DENSITY

ESTIMATION

1

Historical
Background

1.1. Moses as Statistician

A major difference in the orientation of probabilists and statisticians is
the fact that the latter must deal with data from the real world. Realizing the
basic nonstationarity of most interesting data-generating systems, the statistician is frequently as much concerned with the parsimonious representation
of a particular data set as he is with the inferences which can be made from
that data set about the larger population from which the data was drawn.
(In many cases, it makes little sense to talk about a larger population than
the observed data set unless one believes in parallel worlds.)

Let us consider an early statistical investigation—namely, Moses's census
of 1490 B.C. (in Numbers 1 and 2) given below in Table 1.1.

Table 1.1

Tribe	Number of militarily fit males
Judah	74,600
Issachar	54,400
Zebulun	57,400
Reuben	46,500
Simeon	59,300
Gad	45,650
Ephraim ⎱ Joseph	40,500
Manasseh ⎰	32,200
Benjamin	35,400
Dan	62,700
Asher	41,500
Naphtali	53,400
Levi	—
TOTAL	603,550

The tribe of Levi was not included in this or any subsequent census, owing to special priestly exemptions from the misfortunes which inevitably accrue to anyone who gets his name on a governmental roll. A moment's reflection on the table quickly dispels any notions as to a consistent David versus Goliath pattern in Israeli military history. No power in the ancient world could field an army in a Palestinian campaign which could overwhelm a united Israeli home guard by sheer weight of numbers.

Table 1.1 does not appear to have many stochastic attributes. Of course, it could have been used to answer questions, such as "What is the probability that if 20 militiamen are selected at random from the Hebrew host, none will be from the tribe of Judah?" The table is supposedly an exhaustive numbering of eleven tribes; it is not meant to be a random sample from a larger population. And yet, in a sense, the data in Table 1.1 contain much of the material necessary for the construction of a histogram. Two essential characteristics for such a construction are lacking in the data. First, and most important, a proper histogram should have as its domain intervals of the reals (or of R^n). As long as indexing is by qualitative attributes, we are stuck with simply a multinomial distribution. This prevents a natural inferential pooling of adjacent intervals. For example, there is likely to be more relationship between the number of fighters between the heights of 60 and 61 inches and the number of fighters between the heights of 61 and 62 inches than there is between the number of fighters in the two southern tribes of Judah and Benjamin.

Second, no attempt has been made to normalize the table. The ancient statistician failed to note, for example, that the tribe of Issachar contributed 9 percent of the total army. We can only conjecture as to Moses's "feel" about the relative numbers of the host via such things as subsequent battle dispositions.

1.2. The Haberdasher and the "Histogram"

Perhaps one can gain some insight as to the trauma of the perception of the real number system when it is realized that the world had to wait another 3,150 years after the census of Moses for the construction of something approaching a proper histogram. In 1538 A.D., owing to a concern with decreases in the English population caused by the plague, Henry VIII had ordered parish priests of the Church of England to keep a record of christenings, marriages, and deaths [2]. By 1625, monthly and even weekly

Table 1.2

Age interval	Probability of death in interval	Probability of death after interval
0–6	.36	.64
6–16	.24	.40
16–26	.15	.25
26–36	.09	.16
36–46	.06	.10
46–56	.04	.06
56–66	.03	.03
66–76	.02	.01
76–86	.01	0

birth and death lists for the city of London were printed and widely disseminated.

The need for summarizing these mountains of data led John Graunt, a London haberdasher, to present a paper to the Royal Society, in 1661, on the "bills of mortality." This work, *Natural and Political Observations on the Bills of Mortality*, was published as a book in 1662. In that year, on the basis of Graunt's study and on the recommendation of Charles II, the members of the Royal Society enrolled Graunt as a member. (Graunt was dropped from the rostrum five years later, perhaps because he exhibited the anti-intellectual attributes of being (1) a small businessman, (2) a Roman Catholic, and (3) a statistician.) Graunt's entire study is too lengthy to dwell on here. We simply show in Table 1.2 [16, p. 22] his near histogram, which attempts to give probabilities that an individual in London will die in a particular age interval.

Much valid criticism may be made of Graunt's work (see, eg., [2] and [16]). For example, he avoided getting into the much more complicated realm of stochastic processes by tacitly assuming first order stationarity. Then, he gave his readers no clear idea what the cut-off age is when going from one age interval to the next. Finally, he did not actually graph his results or normalize the cell probabilities by the interval widths.

The point to be stressed is that Graunt had given the world perhaps its first empirical cumulative distribution function and was but a normalizing step away from exhibiting an empirical probability density function. To have grasped the desirability of these representations, which had passed uncreated for millenia, is surely a contribution of Newtonian propositions.

Again, the statistical (as opposed to the probabilistic) nature of Graunt's work is clear. Unlike his well-connected contemporary, Pascal, Graunt was not concerned with safe and tightly reasoned deductive arguments on idealized games of chance. Instead, he was involved with making sense out of a mass of practical and dirty data. The summarizing and inferential aspects of his work are equally important and inextricably connected.

It is interesting to speculate whether Graunt visualized the effect of letting his age intervals go to zero; i.e., did he have an empirical feel for modeling the London mortality data with a continuous density function? Of course, derivatives had not yet been discovered, but he would have started down this road empirically if he had constructed a second histogram from the same data, but using a different set of age intervals. It appears likely that Graunt's extreme practicality and his lack of mathematical training did not permit him to get this far. Probably, in Graunt's mind his table was an entity *sui generis* and only vaguely a reflection of some underlying continuous model. In a sense, the empirical cumulative distribution function and empirical probability density function appear to antedate the theoretical entities for which they are estimators.

In the years immediately following publication of Graunt's work, a number of investigators employed histogram and histogram-like means of tabulation. These include Petty, Huygens, van Dael, and Halley [16]. However, it seems clear that each of these built on a knowledge of Graunt's work. We are unaware of any discovery of a near histogram independent of Graunt's treatise.

It is not surprising that a number of years passed before successful attempts were made to place Graunt's work on a solid mathematical footing. His approach in its natural generality is an attempt to estimate a continuous probability density function (of unknown functional form) with a function characterized by a finite number of parameters, where the number of the parameters nevertheless increases without limit as the sample size goes to infinity. The full mathematical arsenal required to undertake the rigorization (and improvement) of Graunt's approach would not be available for two hundred years. We shall not dwell here on the several important examples of continuous probability density models proposed by early investigators for specific situations; e.g., DeMoivre's discovery of the normal density function [15]. Our study is concerned with the estimation of probability densities of unknown functional form. There was very little progress in this task between the work of John Graunt and that of Karl Pearson.

A major difficulty of the histogram (developed in the next chapter) is, of course, the large number of characterizing parameters. The last column of Table 1.2 is unity minus a form of the empirical cumulative distribution function. It requires eight parameters for its description. Should we decide to decrease the interval width, the number of parameters will increase accordingly. In the absence of a computer, it is desirable to attempt to characterize a distribution by a small number of parameters. A popular solution for over one hundred years has been the assumption that every continuous distribution is either normal or nearly normal. Such an optimistic assumption of normality is still common. However, already in the late 1800s, Karl Pearson, that great enemy of "myth" and coiner of the term "histogram" [12], had by extensive examination of data sets perceived that a global assumption of normality was absurd. What was required was the creation of a class of probability densities which possess the properties of fitting a large number of data sets, while having a small number of characterizing parameters.

1.3. The Pearson Distributions

In 1895 Karl Pearson proposed a family of distributions which (as it turned out) included many of the currently more common univariate probability densities as members. "We have, then, reached this point: that to deal effectively with statistics we require generalized probability curves which include the factors of skewness and range" [13, p. 58]. For a thorough discussion of the subject, we refer the reader to Pearson's original work or Johnson's updated and extensively revised version of Elderton's *Frequency Curves and Correlation* [3]. More concise treatments are given by Kendall and Stuart [8], Fisher [4], and Ord [11]. This family of densities is developed, according to the tastes of the time, with a differential equation model.

The hypergeometric distribution is one of the most important in the work of turn-of-the-century physicists and probabilists. We recall that this "sampling without replacement" distribution is the true underlying distribution in a number of situations, where for purposes of mathematical tractability we assume an asymptotically valid approximation such as the binomial or its limit the Gaussian.

Let X be a random variable which the number of black balls drawn in a sample of size n from an urn with Np black balls and $N(1 - p)$ white balls. Then, following the argument of Kendall and Stuart [8],

$$f_x = P[X = x] = \frac{\binom{Np}{x}\binom{N(1 - p)}{n - x}}{\binom{N}{n}}$$

and

$$\frac{\Delta f_x}{f_x} = \frac{P(X = x) - P(X = x - 1)}{P(X = x)} = 1 - \frac{P(X = x - 1)}{P(X = x)}$$

$$= 1 - \frac{x(x-1)!(Np-x)!(N(1-p)-n+x)(N(1-p)-n+x-1)!(n-x)!}{(x-1)!(Np-x+1)(Np-x)!(N(1-p)-n+x-1)!(n-x+1)(n-x)!}$$

$$= 1 - \frac{x(N(1-p)-n+x)}{(Np-x+1)(n-x+1)}$$

$$= \frac{x - (n+1)(Np+1)/(N+2)}{-(n+1)(Np+1)/(N+2) + (Np+n+2)/(N+2)x - (N+2)^{-1}x^2}$$

$$= \frac{(x-a)}{b_0 + b_1 x + b_2 x^2}.$$

Let us consider the associated differential equation

$$(1) \qquad \frac{d(\log f(x))}{dx} = \frac{x - a}{b_0 + b_1 x + b_2 x^2},$$

where $f(x)$ is the probability density function of X. Rewriting (1), we have

$$(2) \qquad (b_0 + b_1 x + b_2 x^2)\, df = (x - a)f\, dx.$$

Multiplying both sides of (2) by x^n and integrating the left-hand side by parts over the range of x, we have

$$(3) \quad (b_0 + b_1 x + b_2 x^2)x^n f \Big|_{x=\ell_1}^{x=\ell_2} - \int_{\ell_1}^{\ell_2} [nb_0 x^{n-1} + (n+1)b_1 x^n$$

$$+ (n+2)b_2 x^{n+1}]f\, dx = \int_{\ell_1}^{\ell_2} x^{n+1}f\, dx - a\int_{\ell_1}^{\ell_2} x^n f\, dx.$$

Assuming that the first term vanishes at the extremities, ℓ_1 and ℓ_2, we have the recurrence relationship

$$(4) \qquad -nb_0\mu'_{n-1} - (n+1)b_1\mu'_n - (n+2)b_2\mu'_{n+1} = \mu'_{n+1} - a\mu'_n$$

or

$$(5) \qquad \mu'_{n+1} = [(a - (n+1)b_1)\mu'_n - nb_0\mu'_{n-1}]/[(n+2)b_2 + 1],$$

where $\mu'_n = E[x^n] = \int x^n f\, dx$ is the nth moment of x about the origin. Let us assume that the origin of x has been placed such that $\mu'_1 = 0$, so that $\mu_n = E[(x - \mu'_1)^n] = \mu'_n$. Letting $n = 0, 1, 2, 3$, and solving for a, b_0, b_1, and b_2 in terms of μ_2, μ_3, and μ_4, we have

(6)
$$a = -\mu_3(\mu_4 + 3\mu_2^2)/Q = -\sqrt{\mu_2}\sqrt{\beta_1}(\beta_2 + 3)/Q'$$
$$b_0 = -\mu_2(4\mu_2\mu_4 - 3\mu_3^2)/Q = -\mu_2(4\beta_2 - 3\beta_1)/Q'$$
$$b_1 = -\mu_3(\mu_4 + 3\mu_2^2)/Q = -\sqrt{\mu_2}\sqrt{\beta_1}(\beta_2 + 3)/Q'$$
$$b_2 = [-2\mu_2\mu_4 + 3\mu_3^2 + 6\mu_2^3]/Q = -(2\beta_2 - 3\beta_1 - 6)/Q',$$

where $Q = 10\mu_2\mu_4 - 18\mu_2^3 - 12\mu_3^2$, $Q' = 10\beta_2 - 18 - 12\beta_1$, $\beta_1 = \mu_3^2/\mu_2^3$ and $\beta_2 = \mu_4/\mu_2^2$. We note that, since $a = b_1$, the Pearson System after translation of the data to make $\mu_1' = 0$ is actually a three-parameter family.

The relations in (6) give us a ready means of estimating a, b_0, b_1, and b_2 via the method of moments. Given a random sample $\{x_1, x_2, \ldots, x_n\}$, we might use as approximation to μ_n', $\dfrac{1}{n}\sum x_j^n$, the nth sample moment.

Clearly, the solution of (1) depends on the roots of the denominator $b_0 + b_1x + b_2x^2$. Letting $\varkappa = b_1^2/4b_0b_2$, we can easily characterize whether we have two real roots with opposite signs (Pearson's Type I), two complex roots (Pearson's Type IV), or two real roots with the same sign (Pearson's Type VI), according as $\varkappa < 0$, $0 < \varkappa < 1$ or $\varkappa > 1$ respectively. The special cases when $\varkappa = 0$ or $\varkappa = 1$ will be discussed later. The development below follows the arguments of Elderton and Johnson [3].

Type I ($\varkappa < 0$)

(7)
$$\frac{d\log f}{dx} = \frac{(x - a)}{b_2(x + A_1)(x - A_2)} \qquad \text{where } A_1, A_2 > 0$$

$$= \frac{1}{b_2}\frac{A_1 + a}{A_1 + A_2}\frac{1}{x + A_1} + \frac{1}{b_2}\frac{A_2 - a}{A_1 + A_2}\frac{1}{x - A_2},$$

giving

$$\log f = \log c + \frac{A_1 + a}{b_2(A_1 + A_2)}\log(x + A_1) + \frac{1}{b_2}\frac{A_2 - a}{A_1 + A_2}\log(A_2 - x)$$

or

(8)
$$f(x) = c(x + A_1)^{(A_1 + a)/b_2(A_1 + A_2)}(A_2 - x)^{(A_2 - a)/b_2(A_1 + A_2)},$$

where c is chosen so that $\int f(x)\,dx = 1$ and $-A_1 < x < A_2$.

This is a beta distribution (of the first kind). We are more familiar with the standard form

(9) $$f(x) = \frac{1}{B(\alpha, \beta)}x^{\alpha - 1}(1 - x)^{\beta - 1}; \alpha, \beta > 0, 0 < x < 1 \text{ and } B(\alpha, \beta) = \frac{\Gamma(\alpha)\Gamma(\beta)}{\Gamma(\alpha + \beta)}.$$

The applications of this distribution include its use as the natural conjugate prior for the binomial distribution. We shall shortly note the relation between Pearson's Type I and his Type VI distributions.

Type IV $(0 < \varkappa < 1)$

(10) $\qquad \dfrac{d \log f}{dx} = \dfrac{(x - a)}{b_2[(x + c)^2 + d^2]}$

$\qquad\qquad = \dfrac{2(x + c)}{2b_2[(x + c)^2 + d^2]} - \dfrac{(c + a)}{b_2} \dfrac{1}{(x + c)^2 + d^2},$

giving

$\qquad \log f = \log k + \dfrac{1}{2b_2} \log[(x + c)^2 + d^2] - \dfrac{c + a}{db_2} \arctan\left(\dfrac{x + c}{d}\right)$

or

(11) $\qquad f(x) = k[(x + c)^2 + d^2]^{1/(2b_2)} \exp\left[-\dfrac{c + a}{db_2} \arctan\left(\dfrac{x + c}{d}\right)\right]$

where $-\infty < x < \infty$ and k is chosen so that $\int f(x)\, dx = 1$.

This distribution is mainly of historical interest, as it is seldom used as a model in data analysis.

Type VI $(\varkappa > 1)$

(12) $\qquad \dfrac{d \log f(x)}{dx} = \dfrac{(x - a)}{b_2(x - A_1)(x - A_2)}$

(where the real valued A_1 and A_2 have the same sign)

$\qquad\qquad = \dfrac{1}{b_2} \dfrac{A_1 - a}{A_1 - A_2} \dfrac{1}{x - A_1} - \dfrac{1}{b_2} \dfrac{A_2 - a}{A_1 - A_2} \dfrac{1}{x - A_2},$

giving

$\qquad \log f = \log c + \dfrac{A_1 - a}{b_2(A_1 - A_2)} \log(x - A_1) - \dfrac{1}{b_2} \dfrac{A_2 - a}{A_1 - A_2} \log(x - A_2)$

or

(13) $\qquad f = c(x - A_1)^{(A_1 - a)/b_2(A_1 - A_2)}(x - A_2)^{-(A_2 - a)/b_2(A_1 - A_2)},$

where $\operatorname{sgn}(a)\, x > \min(|A_1|, |A_2|)$ and c is chosen so that $\int f(x)\, dx = 1$.

This is a beta-distribution of the second kind and is more commonly seen in the standard form

(14) $\qquad f(x) = \dfrac{1}{B(\alpha, \beta)} \dfrac{x^{\alpha - 1}}{(x + 1)^{\alpha + \beta}}, \qquad$ where $\alpha, \beta > 0$ and $x \geq 0$.

Perhaps the most important use of this distribution is in analyses of variance, because the F distribution is a beta distribution of the second kind. Because of the easy transformation of a beta variable of the first or second kind into a beta variable of the other kind, the existence of both Type I and Type VI distributions in the Pearsonian system provides for a kind of redundancy.

If we are to use sample moments in (6) to estimate a, b_0, b_1, and b_2, then we really need only the three members already considered. $\varkappa = \pm\infty$, 0 or 1, the cases not yet examined, will occur very rarely and only as a result of truncation of the decimal representations of the data values. This is reasonably unsettling, since it shows that a direct hammer-and-tongs method of moments on the data will give us only one of two forms of the beta distribution or the rather unwieldy density in (11). For the Pearson family to be worthwhile, it is clear that there must be other distributions of interest for some of the values $\varkappa = \pm\infty$, 0, 1, and that we will not be able to rely simply on data-oriented estimation of a, b_0, b_1, and b_2. Let us consider briefly these important "transition" cases.

Type III ($\varkappa = \infty$)

Let us consider the case where $\varkappa = \infty$. This can occur only if $b_2 = 0$. (Note that $b_0 = 0$ is an impossibility, for then we must have $4\mu_2\mu_4 - 3\mu_3^2 = 0$. But by the Cauchy–Schwarz Inequality, $\mu_2\mu_4 \geq \mu_3^2$. Hence the above equality is impossible). We have, then,

$$(15) \qquad \frac{d \log f}{dx} = \frac{x - a}{b_0 + b_1 x} = \frac{1}{b_1} \frac{x - a}{x + b_0/b_1},$$

giving

$$\log f(x) = \log c + \frac{1}{b_1} x - \frac{ab_1 + b_0}{b_1^2} \log(x + b_0/b_1)$$

or

$$(16) \qquad f(x) = ce^{x/b_1}(x + b_0/b_1)^{-(ab_1 + b_0)/b_1^2}.$$

Changing the origin, we have more concisely

$$(17) \qquad f(x) = c_0 e^{-\gamma x}\left(1 + \frac{x}{c_1}\right)^{\gamma c_1},$$

where $x > -c_1$, $\gamma = 2\mu_2/\mu_3$, $c_1 = \dfrac{2\mu_2^2}{\mu_3} - \dfrac{\mu_3}{2\mu_2}$ and c_0 is chosen so that $\int f(x)\,dx = 1$.

Of course, by shifting the origin to 0 and changing scale via the transformation $x = zc_1 - c_1$, we can obtain the standard gamma distribution

$$(18) \quad f(z) = \frac{1}{\Gamma(\alpha)\beta^\alpha} z^{\alpha-1}e^{-z/\beta}, \qquad \text{where } \alpha = \gamma c_1 + 1 > 0, \quad \beta = 1/(\gamma c_1) > 0.$$

Let us consider the case where $\varkappa = 0$

(19) $$\frac{d \log f}{dx} = \frac{x}{b_0 + b_2 x^2}.$$

Then,

(20) $$d(\log f) = \frac{1}{2b_2} \frac{d(b_0 + b_2 x^2)}{b_0 + b_2 x^2},$$

giving

(21) $$f = c[b_0 + b_2 x^2]^{1/(2b_2)}$$
$$= c'[1 + (b_2/b_0)x^2]^{1/(2b_2)}.$$

Now, from (6) we recall that

$$b_2/b_0 = (2\beta_2 - 6)/(4\mu_2\beta_2).$$

Thus the coefficient of the x^2 term in (21) is negative or positive, depending on whether $\beta_2 < 3$ or $\beta_2 > 3$.

Type II ($\varkappa = 0, \beta_2 < 3$)

(22) $$f = c'[1 - |b_2/b_0|x^2]^{1/(2b_2)},$$

where $-\sqrt{|b_0/b_2|} < x < \sqrt{|b_0/b_2|}$ and c' is chosen so that $\int f(x)\,dx = 1$.

Type VII ($\varkappa = 0, \beta_2 > 3$)

(23) $$f = c'[1 + (b_2/b_0)x^2]^{1/(2b_2)},$$

where (since the exponent $1/(2b_2) < -1$) $-\infty < x < \infty$ and c' is chosen so that $\int f(x)\,dx = 1$.

Gaussian Distribution ($\varkappa = 0, b_1 = b_2 = 0$)

One further case should be considered when $\varkappa = 0$, namely, that in which $b_1 = b_2 = 0$

(24) $$\frac{d \log f}{dx} = \frac{x}{b_0}.$$

From (6) it can be seen that under these conditions b_0 is negative. Thus,

$$\log f = \log c - \frac{x^2}{2|b_0|}.$$

(25) $$f(x) = c \exp\left[-\frac{x^2}{2|b_0|}\right]$$

where $-\infty < x < \infty$ and c is chosen so that $\int f(x)\,dx = 1$.

Type V ($\varkappa = 1$)

When $\varkappa = 1$, $b_0 + b_1 x + b_2 x^2$ has a repeated real root at $-\dfrac{b_1}{2b_2}$

(26)
$$\frac{d \log f}{dx} = \frac{1}{b_2} \frac{x - a}{(x + b_1/2b_2)^2},$$

giving

$$\log f = \log c + \frac{1}{b_2} \log\left(x + \frac{b_1}{2b_2}\right) + \frac{2ab_2 + b_1}{2b_2^2} \frac{1}{(x + b_1/2b_2)}$$

$$= \log c + \log(x + b_1/2b_2)^{1/b_2} + \log\left[\exp\left\{\frac{a + b_1/2b_2}{b_2(x + b_1/2b_2)}\right\}\right],$$

so

(27)
$$f(x) = c(x + b_1/2b_2)^{1/b_2} \exp\left[\frac{a + b_1/2b_2}{b_2(x + b_1/2b_2)}\right]$$

or, letting $x = z - \dfrac{b_1}{b_2}$,

(28)
$$f(z) = cz^{1/b_2} \exp \frac{a + b_1/2b_2}{b_2 z}$$

$$= cz^{-\alpha} e^{-\beta/z},$$

where $\alpha = -\dfrac{1}{b_2}$, $\beta = -\dfrac{a + b_1/2b_2}{b_2}$, $z > 0$ and c is chosen so that $\int f(z)\, dz = 1$.

"Student's" Discovery of the t Distribution

The most famous result in statistics which came about by using the method of moments with the Pearson family appears in a 1908 paper of W. S. Gosset (pseudonym "Student") [6]. Although before Gosset it was well known that the distribution of the sample mean $\bar{x}$ of independent observations from a Gaussian distribution with mean μ and variance σ^2 is Gaussian with mean μ and variance $\dfrac{\sigma^2}{n}$ where n is the sample size, it was unclear what sort of statement could be made when σ^2 was unknown. The usual procedure was to substitute the sample variance s^2 for σ^2; i.e., it was assumed that $\bar{x}$ was Gaussian with mean μ and variance $\dfrac{s^2}{n}$. Gosset has empirically observed that the distribution of $t = \sqrt{n}(\bar{x} - \mu)/s$ is close to the Gaussian for n large, but that for n small the tails of the t distribution were heavier than those of the Gaussian. We sketch below "Student's" derivation of the t distribution.

First, Gosset determined the first four population moments of

$$s^2 = \frac{1}{n-1} \sum_{j=1}^{n} (x_j - \bar{x})^2$$

where the $\{x_j\}$ were assumed to be Gaussian with mean μ and variance σ^2. He found that he obtained a, b_0, b_1 and b_2 values which corresponded in the Pearson family to a Type III curve with origin at 0, namely,

(29) $$f(s^2) = c(s^2)^{(n-3)/2} e^{-ns^2/2\sigma^2}.$$

Next, Gosset found that s and $(\bar{x} - \mu)$ were uncorrelated. Then, assuming that this implied independence of s and $(\bar{x} - \mu)$ (as it turns out is true in this case), he proceeded to derive the distribution of $\sqrt{n}(x - \mu)/s = t$, obtaining

(30) $$f(t) = \frac{\Gamma\left(\dfrac{n}{2}\right)}{\sqrt{(n-1)\pi}\; \Gamma\left(\dfrac{n-1}{2}\right)} \frac{1}{\left(1 + \dfrac{t^2}{n-1}\right)^{n/2}}.$$

It is interesting to note that Gossett actually used the method of moments to obtain the distribution of s^2—not to obtain the distribution of t. The last step he carried out in a manner similar to that usually employed in a contemporary elementary statistics course, where one starts out with the joint density of a x^2 variate and an independent Gaussian variate, introduces t via a transformation, and integrates out the dummy variable.

We note that the t distribution is of Type VII. Had "Student" employed the method of moments on data to infer the distributions of either s^2 or t, he would not have landed in the Type III or Type VII sets—these being of measure zero in the parameter space of (a, b_0, b_1, b_2) as estimated by the method of moments.

This example shows one reason the Pearsonian approach is seldom used at present. If an exact distribution of a measurable function of random variables is to be found, we must know the underlying distribution of the original random variables. We then compute the first four moments of the function and search for the distribution type in the Pearsonian system. (This may be more difficult than finding the distribution directly.) When we finish our work, we have no assurance that the true distribution really is a member of the Pearsonian system. Note that multimodal distributions, for example, are excluded from the system.

Finally, if we seek only an approximation to the underlying distribution for a particular data set, we realize we will land in the parameter sets corresponding to Types I, IV, and VI—a system neither particularly tractable nor flexible. Having made the above comments, we must remember that the most common distributions dealt with in contemporary statistics are members of the Pearsonian family. It is simply that Pearson's approach is rather too ad hoc for theoretical derivation, rather too impractical for ad hoc density estimation.

If the development of statistics proceeded by a sequence of steady Teutonic increments, one might suppose that following Pearson's breakthrough, the main channel of statistics would have proceeded from his work. Accordingly, we might expect to have seen a succession of important generalizations of Pearson's family. Indeed there were a number of such papers, e.g., [7], [10], [19]. However, these studies did not actually push toward the natural goal, namely, the creation of practical algorithms which enable the stable and consistent estimation of probability densities in very general classes. (In a real sense, the subsequent studies of the foremost investigator of probability distributions, Norman Lloyd Johnson, may be said to be motivated in part by Pearson's work.)

Instead, R. A. Fisher appeared on the scene with the concept of maximum likelihood estimation [4] and deflected the thrust and direction of the Pearsonian methodology. The really important concept in Pearson's approach is the desirability in many situations of leaving the a priori functional form of the unknown probability density as unspecified as possible. Unfortunately, the Pearson–Fisher controversy was fought out on the line of the method of moments versus maximum likelihood. The Fisherian victory was nearly complete.

1.4. Density Estimation by Classical Maximum Likelihood

Although over half a century old, maximum likelihood is still the most used of any estimation technique. To motivate this procedure, let us consider a probability density of a random variable x characterized by a real parameter θ:

(31) $$f(x; \theta) = f(x|\theta),$$

where $$-\infty < x < +\infty$$
and $$a \leq \theta \leq b.$$

We have a random sample (collection of n identically and independently distributed random variables) of x's $\{x_1, x_2, \ldots, x_n\}$. Let us suppose we have some knowledge of the true value of θ, which can be characterized by a *prior probability density* $p(\theta)$. The joint density of $x = \{x_1, x_2, \ldots, x_n\}$ and θ is given by

$$(32) \qquad h(x, \theta) = \prod_{j=1}^{n} f(x_j|\theta)p(\theta) = f_n(x|\theta)p(\theta),$$

where $f_n(x|\theta)$ is the *likelihood function*. (It was pointed out to the authors by Salomon Bochner that Fisher was apparently the first researcher to exploit the fact that under the conditions given above, $f_n(x|\theta) = \prod f(x_j|\theta)$.) The marginal density of x is given by

$$(33) \qquad \ell(x) = \int_a^b f_n(x|\theta)p(\theta)\, d\theta.$$

The conditional density of θ given x (i.e., the *posterior density* of θ) is then

$$(34) \qquad g(\theta|x) = \frac{h(x, \theta)}{\ell(x)} = \frac{f_n(x|\theta)p(\theta)}{\int f_n(x|\theta')p(\theta')\, d\theta'}.$$

This is simply a version of Bayes's Theorem. To obtain a good estimate of θ we might use some measure of centrality of the posterior distribution of θ. For example, if we attempt to minimize

$$(35) \qquad E[(\tilde{\theta}(x) - \theta)^2] = \int_x \int_\theta [\tilde{\theta}(x) - \theta]^2 p(\theta) \prod_{j=1}^{n} f(x_j|\theta)\, dx_j\, d\theta$$

we may do so by selecting for each x that value which minimizes

$$(36) \qquad \int_\theta [\tilde{\theta}(x) - \theta]^2 \prod_{j=1}^{n} f(x_j|\theta)p(\theta)\, d\theta.$$

Now, this may be obtained by differentiating with respect to the real valued $\tilde{\theta}(x)$ and setting the derivative equal to zero to give

$$(37) \qquad \tilde{\theta}(x) = \int \theta g(\theta|x)\, d\theta$$

or, simply the mean of the posterior distribution $g(\theta|x)$.

Alternatively, we might seek that value of θ such that

$$(38) \qquad \int_a^{\theta_{me}} g(\theta|x)\, d\theta = \int_{\theta_{me}}^{b} g(\theta|x)\, d\theta = .5.$$

That is, we could use the posterior median as an estimate for θ.

As a third possibility we might seek a value of θ which maximizes the posterior density function of θ. If we are fortunate, the posterior mode will be unique.

Let us consider the case where we have only very vague notions as to the true value of θ. We only know that θ cannot possibly be smaller than some a or larger than some b. Then we might decide to use as the prior distribution for θ the uniform distribution on the interval $[a, b]$, namely,

(39)
$$p(\theta) = \frac{1}{b - a} \quad \text{if} \quad a \le \theta \le b$$
$$= 0, \qquad \text{otherwise.}$$

Then,

(40)
$$g(\theta|x) = \frac{f_n(x|\theta) \dfrac{1}{(b - a)}}{\displaystyle\int_a^b f_n(x|\theta') \frac{1}{b - a} \, d\theta'},$$
$$\text{if } a \le \theta \le b$$
$$= \frac{f_n(x|\theta)}{\displaystyle\int_a^b f_n(x|\theta') \, d\theta'}$$
$$= 0, \qquad \text{otherwise.}$$

Thus, to obtain a value of θ which maximizes $g(\theta|x)$ we need only find a value of θ which maximizes the likelihood

(41)
$$f_n(x|\theta) = \prod_{j=1}^{n} f(x_j|\theta).$$

Such an estimator is called a *maximum likelihood estimator for* θ. We note that in comparison to the other procedures mentioned for obtaining an estimator for θ, the maximum likelihood approach has the advantage of (relative) computational simplicity. However, we note that its use is equivalent to going through a Bayesian argument using a rather noninformative prior. In fact, we have in effect used Bayes's Axiom; i.e., in the absence of information to the contrary, we have assumed all values of θ in $[a, b]$ to be equally likely. Such an assumption is certainly open to question.

Moreover, we have, in choosing maximum likelihood, opted for using the mode of the posterior distribution $g(\theta|x)$ as an estimate for θ based on the data. The location of the global maximum of a stochastic curve is generally much more unstable than either the mean or the median of the curve. Consequently, we might expect maximum likelihood estimators frequently to behave poorly for small samples. Nonetheless, under very general conditions they have excellent large sample properties, as we shall show below.

Following Wilks [18], let us consider the case where the continuous probability density function $f(.|\theta)$ is regular with respect to its first θ derivative in the parameter space Θ, i.e.,

(42) $E\left[\dfrac{\partial \log f(x|\theta)}{\partial \theta}\right] = \dfrac{\partial}{\partial \theta} \int \log f(x|\theta) f(x|\theta)\, dx = \dfrac{\partial}{\partial \theta} \int f(x|\theta)\, dx = 0.$

Let us assume that for any given sample $x = (x_1, x_2, \ldots, x_n)$ equation (43) below has a unique solution.

(43) $\left[\dfrac{\partial}{\partial \theta} \log f_n(x|\theta)\right]_{\hat{\theta}_n} = \sum_{j=1}^{n}\left[\dfrac{\partial}{\partial \theta} \log f(x_j|\theta)\right]_{\hat{\theta}_n} = 0.$

Now, let us suppose our sample $(x_1, x_2, \ldots, x_n)$ has come from $f(.|\theta_0)$, i.e., the actual value of θ is the point $\theta_0 \in \Theta$. We wish to examine the stochastic convergence of $\hat{\theta}_n$ to θ_0. We shall assume that

$$B^2(\theta,\theta) = \text{var}\left[\frac{\partial \log f(x|\theta)}{\partial \theta}\right] = \int \left(\frac{\partial \log f(x|\theta)}{\partial \theta}\right)^2 f(x|\theta)\, dx < \infty.$$

Let us examine

(44) $H(\theta_0, \theta) = \int [\log f(x|\theta)] f(x|\theta_0)\, dx.$

Considering the second difference of $H(\theta_0, .)$ about θ_0, we have (taking care that $[\theta_0 - h, \theta_0 + h]$ is in Θ)

(45) $\Delta^2_{\theta_0,h} H(\theta_0, \theta)$

$\qquad = \int [\log f(x|\theta_0 + h)] f(x|\theta_0)\, dx$

$\qquad\quad + \int [\log f(x|\theta_0 - h)] f(x|\theta_0)\, dx - 2 \int [\log f(x|\theta_0)] f(x|\theta_0)\, dx$

$\qquad = \int \log\left[\dfrac{f(x|\theta_0 + h)}{f(x|\theta_0)}\right] f(x|\theta_0)\, dx + \int \log\left[\dfrac{f(x|\theta_0 - h)}{f(x|\theta_0)}\right] f(x|\theta_0)\, dx$

$\qquad < \log \int \dfrac{f(x|\theta_0 + h)}{f(x|\theta_0)} f(x|\theta_0)\, dx + \log \int \dfrac{f(x|\theta_0 - h)}{f(x|\theta_0)} f(x|\theta_0)\, dx$

(by the strict concavity of the logarithm and Jensen's Inequality)

$\qquad = \log 1 + \log 1.$

Thus,

$$\Delta^2_{\theta_0,h} H(\theta_0, \theta) < 0.$$

But then

$$A(\theta_0, \theta) = \int \left[\frac{\partial \log f(x|\theta)}{\partial \theta} \right] f(x|\theta_0) \, dx$$

is strictly decreasing in a neighborhood $(\theta_0 - \delta, \theta_0 + \delta) \in \Theta$ about θ_0.

Now, $\dfrac{1}{n} \dfrac{\partial}{\partial \theta} \log f_n(x|\theta) = \dfrac{1}{n} \sum \dfrac{\partial}{\partial \theta} \log f(x_j|\theta)$ is the sample mean of a sample

of size n of the random variable $Y = \dfrac{\partial}{\partial \theta} \log f(x|\theta)$. Since

$$E[Y] = \int \frac{\partial}{\partial \theta} [\log f(x|\theta)] \, f(x|\theta_0) = A(\theta_0, \theta),$$

by the Strong Law of Large Numbers, we know that $\dfrac{1}{n} \dfrac{\partial}{\partial \theta} \log f_n(x|\theta)$ converges

almost surely to $A(\theta_0, \theta)$. Moreover, by (42) we have $A(\theta_0, \theta_0) = 0$. Thus, for any $\epsilon > 0$, $0 < \delta' < \delta$, there may be found an $n(\delta', \epsilon)$, such that probability exceeds $1 - \epsilon$ that the following inequalities hold for all $n > n(\delta', \epsilon)$:

$$\frac{1}{n} \frac{\partial}{\partial \theta} \log f_n(x|\theta) > 0 \quad \text{if} \quad \theta = \theta_0 - \delta'$$

$$\frac{1}{n} \frac{\partial}{\partial \theta} \log f_n(x|\theta) < 0 \quad \text{if} \quad \theta = \theta_0 + \delta'.$$

Consequently, if $\dfrac{\partial}{\partial \theta} \log f_n(x|\theta)$ is continuous in θ over $(\theta_0 - \delta', \theta_0 + \delta')$

(46)
$$P\left[\frac{1}{n} \frac{\partial}{\partial \theta} \log f_n(x|\theta) = 0, \right.$$

for some θ in $(\theta_0 \pm \delta')$ for all $n > n(\delta', \epsilon)|\theta_0] > 1 - \epsilon$. But since we have assumed that $\hat{\theta}_n$ uniquely solves (43), we have that the maximum likelihood estimator converges almost surely to θ_0; i.e.,

(47)
$$\lim_{n \to \infty} P[\hat{\theta}_n = \theta_0] = 1.$$

Thus, we have

Theorem 1. Let $(x_1, x_2, \ldots, x_n)$ be a sample from the distribution with probability density function $f(.|\theta)$. Let $f(.|\theta)$ be regular with respect to its

first θ derivative in $\Theta \subset R_1$. Let $\dfrac{\partial}{\partial \theta} \log f(.|\theta)$ be continuous in θ for all

values of $x \in R$, except possibly for a set of probability zero. Suppose (43) has a unique solution, say, $\hat{\theta}_n(x_1, x_2, \ldots, x_n)$, for any n and almost all $(x_1, x_2, \ldots, x_n) \in R_n$. Then the sequence $\{\hat{\theta}_n(x_1, x_2, \ldots, x_n)\}$ converges almost surely to θ_0.

More generally, Wald [17] has considered the case where there may be a multitude of relative maxima of the log likelihood. He has shown that, assuming a number of side conditions, if we take for our estimator a value of $\hat{\theta}_n \in \Theta$, which gives an absolute maximum of the likelihood, then $\hat{\theta}_n$ will converge almost surely to θ_0. We have presented the less general result of Wilks because of its relative brevity. (Simplified and abbreviated versions of Wald's proof of which we are aware tend to be wrong, e.g., [9, pp. 40–41].)

Now, an estimator $\hat{\theta}_n$ for a parameter θ_0 is said to be consistent if $\{\hat{\theta}_n\}$ converges to θ_0 in probability. Since almost sure convergence implies convergence in probability, we note that maximum likelihood estimates are consistent. That consistency is actually a fairly weak property is seen by noting that if $\hat{\theta}_n(x)$ is consistent for θ_0, so also is $\hat{\theta}_n(y)$ where y is the censored sample obtained by throwing away all but every billionth observation. Clearly, consistency is not a sufficient condition for an estimator to be satisfactory. From a theoretical standpoint consistency is not a necessary condition either. Since we will never expect to know precisely the functional form of the probability density, and since the true probability density is probably not precisely stationary during the period (in time, space, etc.) over which the observations were collected, it is unlikely that any estimator, say $\breve{\theta}$, we might select will converge in probability to θ_0. In fact, it is unlikely that θ_0 can itself be well defined. Of course, this sort of argument can be used to invalidate all human activity whatever. As a practical matter formal consistency is a useful condition for an estimator to possess.

Had Fisher only been able to establish the consistency of maximum likelihood estimators, it is unlikely he would have won his contest with Pearson. But he was able to show [4] a much stronger asymptotic property.

To obtain an upper bound on the rate of convergence of an estimator for θ, let us, following Kendall and Stuart [9], consider

$$(48) \qquad \int \cdots \int f_n(x|\theta) \prod dx_j = 1.$$

Assuming (48) is twice differentiable under the integral sign, we have differentiating once

$$(49) \qquad 0 = \int \cdots \int \frac{\partial f_n}{\partial \theta} \prod dx_j = \int \cdots \int \left(\frac{1}{f_n} \frac{\partial f_n}{\partial \theta} \right) f_n \prod dx_j$$

$$= \int \cdots \int \frac{\partial \log f_n}{\partial \theta} f_n \prod dx_j.$$

Differentiating again

$$\int \cdots \int \left[f_n \frac{\partial}{\partial \theta} \left(\frac{1}{f_n} \frac{\partial f_n}{\partial \theta} \right) + \left(\frac{\partial f}{\partial \theta} \right)^2 \frac{1}{f_n} \right] \prod dx_j$$

$$= \int \cdots \int \left[\frac{\partial^2 \log f_n}{\partial \theta^2} + \left(\frac{1}{f_n} \frac{\partial f_n}{\partial \theta} \right)^2 \right] f_n \prod dx_j = 0,$$

which gives

(50)
$$E\left[\frac{\partial^2 \log f_n}{\partial \theta^2} \right] = -E\left[\left(\frac{\partial \log f_n}{\partial \theta} \right)^2 \right].$$

Now, if we have an unbiased estimator for θ, say $\breve{\theta}$, then,

(51)
$$\int \cdots \int (\breve{\theta}(x_n) - \theta) f_n(x|\theta) \prod dx_j = 0.$$

Differentiating with respect to θ

(52)
$$\int \cdots \int (\breve{\theta} - \theta) \frac{\partial \log f_n}{\partial \theta} f_n \prod dx_j = 1$$

or

$$E\left[(\breve{\theta} - \theta) \left(\frac{\partial \log f_n}{\partial \theta} \right) \right] = 1,$$

giving by the Cauchy–Schwarz Inequality:

(53)
$$E[(\breve{\theta} - \theta)^2] \geq \frac{1}{E\left[\left(\frac{\partial \log f_n}{\partial \theta} \right)^2 \right]} = \frac{-1}{E\left[\frac{\partial^2 \log f_n}{\partial \theta^2} \right]}.$$

The Cramér-Rao Inequality in (53), shows that no unbiased estimator can have smaller mean square error than $\dfrac{-1}{E\left[\dfrac{\partial^2 \log f_n}{\partial \theta^2} \right]}$. Since the observations are independent, this can easily be shown to be, simply $\dfrac{-1}{nE\left[\left(\dfrac{\partial^2 \log f(x|\theta)}{\partial \theta^2} \right) \right]}$.

Theorem 2. Let the first two derivatives of $f_n(x_n|\theta)$ with respect to θ exist in an interval about the true value θ_0. Furthermore, let

$$E\left[\frac{\partial \log f_n}{\partial \theta} \right] = 0 \quad \text{and}$$

$$B_n^2(\theta) = E\left[\left(\frac{\partial \log f_n}{\partial \theta} \right)^2 \right] = -E\left[\frac{\partial^2 \log f_n}{\partial \theta^2} \right] = -nE\left[\frac{\partial^2 \log f}{\partial \theta^2} \right]$$

be nonzero for all θ in the interval. Then the maximum likelihood estimator $\hat{\theta}_n$ is asymptotically Gaussian with mean θ_0 and variance equal to the Cramer-Rao Lower Bound $1/B_n^2(\theta_0)$.

Proof

Expanding $\left(\dfrac{\partial \log f_n}{\partial \theta}\right)_{\hat{\theta}_n}$ about θ_0, we have

$$(54) \qquad \left(\frac{\partial \log f_n}{\partial \theta}\right)_{\hat{\theta}_n} = \left(\frac{\partial \log f_n}{\partial \theta}\right)_{\theta_0} + (\hat{\theta}_n - \theta_0)\left(\frac{\partial^2 \log f_n}{\partial \theta^2}\right)_{\theta_n'},$$

where θ_n' is between θ_0 and $\hat{\theta}_n$. Since the left-hand side of (54) is zero,

$$(55) \qquad (\hat{\theta}_n - \theta_0) = -\frac{\left(\dfrac{\partial \log f_n}{\partial \theta}\right)_{\theta_0}}{\left(\dfrac{\partial^2 \log f_n}{\partial \theta^2}\right)_{\theta_n'}}, \qquad \text{where } \theta_n' \text{ is between } \theta_0 \text{ and } \hat{\theta}_n,$$

or, rewriting

$$(56) \qquad (\hat{\theta}_n - \theta_0)B_n(\theta_0) = \frac{\left(\dfrac{\partial \log f_n}{\partial \theta}\right)_{\theta_0}\bigg/ B_n(\theta_0)}{\left(\dfrac{\partial^2 \log f_n}{\partial \theta^2}\right)_{\theta_n'}\bigg/ -B_n^2(\theta_0)}$$

$$= \frac{\displaystyle\sum_{j=1}^{n}\left(\dfrac{\partial \log f(x_j)}{\partial \theta}\right)_{\theta_0}\bigg/ B_n(\theta_0)}{\dfrac{1}{n}\displaystyle\sum_{j=1}^{n}\left(\dfrac{\partial^2 \log f(x_j)}{\partial \theta^2}\right)_{\theta_n'}\bigg/ E\left[\left(\dfrac{\partial^2 \log f}{\partial \theta^2}\right)_{\theta_0}\right]}.$$

But we have seen that $\lim\limits_{n\to\infty} P(\hat{\theta}_n = \theta_0) = 1$. Hence, in the limit the denominator on the right-hand side of (56) becomes unity.

Moreover, we note that

$$(57) \qquad \frac{\partial \log f_n}{\partial \theta} = \sum_{j=1}^{n} \frac{\partial \log f(x_j)}{\partial \theta} = \sum_{j=1}^{n} y_j,$$

where the $\{y_j\}$ are independent random variables with mean zero and variance

$$E\left[\left(\frac{\partial \log f(x|\theta)}{\partial \theta}\right)_{\theta_0}^2\right] = \frac{1}{n}B_n^2(\theta_0).$$

Thus, the numerator of the right-hand side of (56) converges in distribution to a Gaussian variate with mean zero and variance 1. Finally, then, $\hat{\theta}_n$

converges in distribution to the Gaussian distribution with mean θ_0 and variance $1/B_n^2(\theta_0)$. ∎

Now, the efficiency of an estimator $\theta_n^{(1)}$ for the scalar parameter θ_0 relative to a second estimator $\theta_n^{(2)}$ may be defined by

(58) $$\text{Eff}[\theta_n^{(1)}|\theta_n(2)] = \frac{E[(\theta_n^{(2)}(x_1, x_2, \ldots, x_n) - \theta_0)^2]}{E[(\theta_n^{(1)}(x_1, x_2, \ldots, x_n) - \theta_0)^2]}.$$

But if a maximum likelihood estimator for θ_0 exists, it asymptotically achieves the Cramér-Rao Lower Bound $1/B_n^2(\theta_0)$ for variance. Thus, if we use as the standard $\theta_n^{(2)}$, a hypothetical estimator which achieves the Cramér-Rao lower variance bound, we have as the asymptotic efficiency of the maximum likelihood estimator 100%. Although he left the rigorous proofs for later researchers (e.g., Cramér [1]), Fisher, in 1922, stated this key result [4] while pointing out in great detail the generally poor efficiencies obtained in the Pearson family if the method of moments is employed. Fisher neatly side-stepped the question of what to do in case one did not know the functional form of the unknown density. He did this by separating the problem of determining the form of the unknown density (in Fisher's terminology, the problem of "specification") from the problem of determining the parameters which characterize a specified density (in Fisher's terminology, the problem of "estimation"). The specification problem was to be solved by extramathematical means: "As regards problems of specification, these are entirely a matter for the practical statistician, for those cases where the qualitative nature of the hypothetical population is known do not involve any problems of this type. In other cases we may know by experience what forms are likely to be suitable, and the adequacy of our choice may be tested *a posteriori*: We must confine ourselves to those forms which we know how to handle, or for which any tables which may be necessary have been constructed."

Months before his death [14] Pearson *seemingly* pointed out the specification-estimation difficulty of classical maximum likelihood. "To my astonishment that method depends on first working out the constants of the frequency curve by the Method of Moments, and then superposing on it, by what Fisher terms the "Method of Maximum Likelihood," a further approximation to obtain, what he holds he will thus get, "more efficient values" of the curve constants." Actually, to take the statement in context, Pearson was objecting to the use of moments estimators as "starters" for an iterative algorithm to obtain maximum likelihood estimators. Apparently, Pearson failed to perceive the major difficulty with parametric maximum likelihood estimation. His arguments represent a losing rear-guard action. Nowhere does he make the point that since maximum likelihood assumes a great deal to be "given," it should surprise no one that it is more efficient than more general procedures *when the actual functional form of the density is that assumed.* But what about the fragility of maximum likelihood when the wrong functional form is assumed?

Fisher's answering diatribe [5] shortly after Pearson's death was unnecessary. By tacitly conceding Fisher's point that the underlying density's functional form is known prior to the analysis of a data set. Pearson had lost the controversy.

Neither in the aforementioned harsh "obituary" of one of Britain's foremost scientific intellects, nor subsequently, did Fisher resolve the specification-estimation difficulty of classical maximum likelihood. The victory of Fisher forced upon statistics a parametric straightjacket which it wears to this day. Although Fisher was fond of pointing out the difficulties of assuming the correct prior distribution, $p(\theta)$, in (32), he did not disdain to make a prodigious leap of faith in his selection of $f(x|\theta)$. In the subsequent chapters of this book we wish to examine techniques whereby we may attack the problem of estimating the density function of x without prior assumptions (other than smoothness) as to its functional form.

References

[1] Cramér, Harald (1946). *Mathematical Methods of Statistics*. Princeton: Princeton University Press.

[2] David, F. N. (1962). *Games, Gods and Gambling*. New York: Hafner.

[3] Elderton, William Palen, and Johnson, Norman Lloyd (1969). *Systems of Frequency Curves*. Cambridge: Cambridge University Press.

[4] Fisher, R. A. (1922). "On the mathematical foundations of theoretical statistics." *Philosophical Transactions of the Royal Society of London, Series A* 222: 309–68.

[5] _____ (1937). "Professor Karl Pearson and the method of moments." *Annals of Eugenics* 7: 303–18.

[6] Gosset, William Sealy (1908). "The probable error of a mean." *Biometrika* 6: 1–25.

[7] Hansmann, G. H. (1934). "On certain non-normal symmetrical frequency distributions." *Biometrika* 26: 129–95.

[8] Kendall, Maurice G., and Stuart, Alan (1958). *The Advanced Theory of Statistics*, Vol. I. New York: Hafner.

[9] _____ (1961). *The Advanced Theory of Statistics*, Vol. II. New York: Hafner.

[10] Mouzon, E. D. (1930). "Equimodal frequency distributions." *Annals of Mathematical Statistics* 1: 137–58.

[11] Ord, J. K. (1972). *Families of Frequency Distributions*. New York: Hafner.

[12] Pearson, E. S. (1965). "Studies in the history of probability and statistics. XIV. Some incidents in the early history of biometry and statistics, 1890–94." *Biometrika* 52: 3–18.

[13] Pearson, Karl (1895). "Contributions to the mathematical theory of evolution. II. Skew variation in homogeneous material." *Philosophical Transactions of the Royal Society of London, Series A* 186; 343–414.

[14] ―――― (1936). "The method of moments and the method of maximum likelihood." *Biometrika* 28: 34–59.

[15] Walker, Helen M. (1929). *Studies in the History of Statistical Method.* Baltimore: Williams and Wilkins.

[16] Westergaard, Harald (1968). *Contributions to the History of Statistics.* New York: Agathon.

[17] Wald, Abraham (1949). "Note on the consistency of the maximum likelihood estimate." *Annals of Mathematical Statistics* 20: 595–601.

[18] Wilks, Samuel S. (1962). *Mathematical Statistics.* New York: John Wiley and Sons.

[19] Zoch, R. T. (1934). "Invariants and covariants of certain frequency curves." *Annals of Mathematical Statistics* 5: 124–35.

2

Some Approaches
to Nonparametric Density
Estimation

2.1. The Normal Distribution as Universal Density

Francis Galton, perhaps more than any other researcher, deserves the title of Prophet of the Normal Distribution: "I know of scarcely anything so apt to impress the imagination as the wonderful form of cosmic order expressed by the 'Law of Frequency of Error.' The law would have been personified by the Greeks and deified, if they had known of it. It reigns with serenity and in complete self-effacement amidst the wildest confusion. The huger the mob and the greater the apparent anarchy, the more perfect is its sway. It is the supreme law of Unreason" [20, p. 66]. Moreover, Galton perceived the heuristics of the Central Limit Theorem: "The (normal) Law of Error finds a footing whenever the individual peculiarities are wholly due to the combined influence of a multitude of 'accidents'" [20, p. 55].

As one form of this theorem, we give Rényi's version [37, p. 223] of Lindeberg's 1922 result [31].

Theorem 1. (Central Limit Theorem). Let $x_1, x_2, \ldots, x_n, \ldots$, be a sequence of independent random variables. Let $E(x_n) = 0$ and $E(x_n^2) < \infty$ for all n. Let $F_n(x) = P(x_n \le x)$.
Let

(1)
$$z_n = \sum_{j=1}^{n} x_j$$

and

(2)
$$s_n^2 = E(z_n^2) = \sum_{j=1}^{n} E(x_j^2).$$

For $\epsilon > 0$, let

(3)
$$B_n^*(\epsilon) = \frac{1}{s_n^2} \sum_{j=1}^{n} \int_{|x| > \epsilon s_n} x^2 \, dF_j(x).$$

Suppose that for every $\epsilon > 0$,

(4)
$$\lim_{n \to \infty} B_n^*(\epsilon) = 0.$$

Then one has uniformly (in x):

(5)
$$\lim_{n \to +\infty} P\left(\frac{z_n}{s_n} < x\right) = N(x),$$

where $N(x) = \dfrac{1}{\sqrt{2\pi}} \displaystyle\int_{-\infty}^{x} e^{-u^2/2} \, du.$

Proof.

Let
$$\|f\| = \sup_{x} |f(x)| \quad \text{for} \quad f \in C(-\infty, \infty),$$

where $C(-\infty, \infty)$ is the set of all bounded continuous functions defined on the real line.

For $F(\cdot)$ any cumulative probability distribution function, let us define the operator A_F on $C(-\infty, \infty)$ by

(6)
$$A_F f(x) = \int_{-\infty}^{\infty} f(x + y) \, dF(y).$$

Clearly, A_F is linear. Also, since no shifted average of f can be greater than the supremum of f,

(7)
$$\|A_F f(x)\| \le \|f(x)\|,$$

i.e., A_F is a contraction operator. And if $G(\cdot)$ is another *cdf*

(8)
$$A_F A_G f(x) = A_G A_F f(x) = \int_{-\infty}^{\infty} \int_{-\infty}^{\infty} f(x + y + z) \, dF(y) \, dG(z)$$
$$= \int_{-\infty}^{\infty} f(x + u) \, dH(u) = A_H f(x), \quad \text{where } H = F * G.$$

Now if $A_1, A_2, \ldots, A_n, B_1, B_2, \ldots, B_n$ are linear contraction operators, since

(9)
$$\prod_{j=1}^{n} A_j - \prod_{\ell=1}^{n} B_\ell = \sum_{k=1}^{n} \left[\left(\prod_{i<k} A_i \right) (A_k - B_k) \prod_{j>k} B_j \right].$$

We have

(10)
$$\|A_1 A_2 \cdots A_n f - B_1 B_2 \cdots B_n f\| \le \sum_{k=1}^{n} \|A_k f - B_k f\|.$$

Let $G_n(x) = P[z_n/s_n \le x]$.

Then,

(11) $$G_n(x) = G_{n1}(x)*G_{n2}(x)* \cdots *G_{nn}(x),$$

where

(12) $$G_{nk}(x) = F_k(xs_n) \quad \text{for} \quad k = 1, 2, \ldots, n$$
$$= P[x_k \le xs_n] = P[x_k/s_n \le x].$$

So

(13) $$\int_{-\infty}^{\infty} x \, dG_{nk}(x) = \int_{-\infty}^{\infty} x \, dF_k(xs_n)$$
$$= \frac{1}{s_n} \int \zeta \, dF_k(\zeta)$$
$$= \frac{1}{s_n} E(x_k) = 0$$

and

(14) $$\int_{-\infty}^{\infty} x^2 \, dG_{nk}(x) = \frac{1}{s_n^2} E(x_k^2).$$

Now,

(15) $$N(x) = N_{n1}(x)*N_{n2}(x)* \cdots *N_{nn}(x),$$

where

(16) $$N_{nk}(x) = N\left(\frac{xs_n}{\sqrt{E(x_k^2)}}\right).$$

But

(17) $$\int_{-\infty}^{\infty} x \, dN_{nk}(x) = 0$$

and

(18) $$\int x^2 \, dN_{nk}(x) = \frac{E(x_k^2)}{s_n^2}.$$

From (10), we have for every $f \in C(-\infty, \infty)$

(19) $$\|A_{G_n}f - A_N f\| \le \sum_{k=1}^{n} \|A_{G_{nk}}f - A_{N_{nk}}f\|.$$

Now, denoting by $C^3(-\infty, \infty)$ the set of functions having the first three derivatives continuous and bounded, then if $f \in C^3(-\infty, \infty)$,

$$(20) \qquad A_{G_{nk}} f = f(x) + \frac{f''(x)E(x_k^2)}{2s_n^2}$$

$$+ \frac{1}{6} \int_{|y| \le \epsilon} y^3 f'''(x + \zeta_1 y) \, dG_{nk}(y)$$

$$+ \frac{1}{2} \int_{|y| > \epsilon} y^2 [f''(x + \zeta_2 y) - f''(x)] \, dG_{nk}(y),$$

where $0 \le \zeta_1 \le 1$ and $0 \le \zeta_2 \le 1$.
Similarly,

$$(21) \qquad A_{N_{nk}} f = f(x) + \frac{f''(x)E(x_k^2)}{2s_n^2} + \frac{1}{6} \int_{|y| \le \epsilon} y^3 f'''(x + \zeta_1 y) \, dN_{nk}(y)$$

$$+ \frac{1}{2} \int_{|y| > \epsilon} y^2 [f''(x + \zeta_2 y) - f''(x)] \, dN_{nk}(y).$$

Thus,

$$(22) \quad \|A_{G_{nk}} f - A_{N_{nk}} f\| \le k_1 \left[\int_{|y| > \epsilon} y^2 \, dG_{nk}(y) + \int_{|y| > \epsilon} y^2 \, dN_{nk}(y) \right]$$

$$+ \frac{k_2 \epsilon}{6} \left[\int_{|y| \le \epsilon} y^2 \, dG_{nk}(y) + \int_{|y| \le \epsilon} y^2 \, dN_{nk}(y) \right],$$

where $\sup|f''(x)| = k_1$ and $\sup|f'''(x)| = k_2$. Now,

$$\frac{k_2 \epsilon}{6} \left[\int_{|y| \le \epsilon} y^2 \, dG_{nk}(y) + \int_{|y| \le} y^2 \, dN_{nk}(y) \right] \le \frac{k_2 \epsilon E(x_k^2)}{3s_n^2};$$

$$\int_{|y| > \epsilon} y^2 \, dG_{nk}(y) = \frac{1}{s_n^2} \int_{|y| > \epsilon s_n} y^2 \, dF_k(y);$$

$$\int_{|y| > \epsilon} y^2 \, dN_{nk}(y) = \frac{E(x_k^2)}{s_n^2} \int_{|u| > \epsilon s_n / D_k} \frac{u^2 e^{-u^2/2} \, du}{\sqrt{2\pi}}.$$

So

$$(23) \quad \|A_{G_n} f - A_N f\| \le \frac{k_2 \epsilon}{3} + \frac{k_1}{2} \left(B_n^*(\epsilon) + \frac{1}{\sqrt{2\pi}} \int_{|u| > \epsilon s_n / \Delta_n} u^2 e^{-u^2/2} \, du \right),$$

where $\Delta_n = \max_{1 \le k \le n} \sqrt{E(x_k^2)}$.

But for each $k \le n$ and every $\epsilon' > 0$

(24) $\qquad \dfrac{E(x_k^2)}{s_n^2} = \dfrac{1}{s_n^2} \displaystyle\int_{-\infty}^{\infty} x^2 \, dF_k(x)$

$$= \dfrac{1}{s_n^2} \int_{|x| > \epsilon' s_n} x^2 \, dF_k(x) + \dfrac{1}{s_n^2} \int_{|x| \le \epsilon' s_n} x^2 \, dF_k(x)$$

$$\le B_n^*(\epsilon') + \epsilon'^2.$$

Hence,

(25) $\qquad\qquad\qquad\qquad \dfrac{\Delta_n^2}{s_n^2} \le \epsilon'^2 + B_n^*(\epsilon').$

So, by (4),

$$\lim_{n \to \infty} \dfrac{\Delta_n}{S_n} \le \epsilon'.$$

Since ϵ' can be chosen arbitrarily small,

$$\lim_{n \to \infty} \dfrac{\Delta_n}{S_n} = 0.$$

Thus, we have uniformly in x

(26) $\qquad\qquad\qquad\qquad \lim_{n \to \infty} A_{G_n} f = A_N f.$

Therefore, we have for each $f \in C^3(-\infty, \infty)$

(27) $\qquad\qquad \lim_{n \to \infty} \displaystyle\int_{-\infty}^{\infty} f(y) \, dG_n(y) = \int_{-\infty}^{\infty} f(y) \, dN(y).$

Now (27) holds for all f in $C^3(-\infty, \infty)$. Consider the following C^3 function:

$$f(\epsilon, x, y) = \begin{cases} 1 & \text{for } -\infty < y < x \\ \left[1 - \left(\dfrac{x-y}{\epsilon} \right)^4 \right]^4 & \text{for } x \le y \le x + \epsilon. \\ 0 & \text{for } x + \epsilon < y \end{cases}$$

Further, by (27) we know that

$$\lim_{n \to \infty} \int_{-\infty}^{\infty} f(\epsilon, x, y) \, dG_n(y) = \int_{-\infty}^{\infty} f(\epsilon, x, y) \, dN(y).$$

But

$$\int_{-\infty}^{\infty} f(\epsilon, x, y) \, dN(y) \le N(x + \epsilon) \quad \text{for all } \epsilon > 0.$$

Hence

$$\lim_{n \to \infty} \sup \, G_n(x) \le N(x + \epsilon).$$

Similarly by looking at $f(\epsilon, x - \epsilon, y)$ we find that

$$\liminf_{n \to \infty} G_n(x) \geq N(x - \epsilon).$$

Thus

$$\lim_{n \to \infty} G_n(x) = N(x). \quad \blacksquare$$

The Central Limit Theorem does not state that a random sample from a population satisfying the Lindeberg conditions will have an approximately normal distribution if the sample size is sufficiently large. It is rather the sample sum, or equivalently the sample mean, which will approach normality as the sample size becomes large. Thus, if one were to obtain average IQ's from say, 1,000 twelfth-grade students chosen at random from all twelfth-grade classes in the United States, it might be reasonable to apply normal theory to obtain a confidence interval for the average IQ of American twelfth-grade students. It would not be reasonable to infer that the population of IQ's of twelfth-grade students is normal. In fact, such IQ populations are typically multimodal.

Yet Galton treated almost any data set as though it had come from a normal distribution. Such practice is common to this day among many applied workers—particularly in the social sciences. Frequently, such an assumption of normality does not lead to an erroneous analysis. There are many reasons why this may be so in any particular data analysis.

For example, the Central Limit Theorem may indeed be driving the data to normality. Let us suppose that each of the observations $\{x_1, x_2, \ldots, x_n\}$ in a data set is the average of N independent and identically distributed random variables $\{y_{i1}, y_{i2}, \ldots, y_{iN}\}$, $(i = 1, 2, \ldots, n)$, having finite mean μ and variance σ^2. Then, for N very large each x will be approximately normal with mean μ and variance σ^2/N. We observe that the size of n has nothing to do in bringing about the approximate normality of the data. It need not be the case that the $\{y_{ij}\}$ be observables—in fact, it is unlikely that one would even know what y is.

Generalizations on the above conditions can be made which will still leave the $\{x_i\}$ normal. For example [10, p. 218], suppose

$$(28) \qquad x = g(y_1, y_2, \ldots, y_N),$$

where g has continuous derivatives of the first and second orders in the neighborhood of $\mu = (\mu_1, \mu_2, \ldots, \mu_N)$, where $\mu_j = E(y_j)$. Then we may write

$$(29) \quad g(y_1, y_2, \ldots, y_N) = g(\mu_1, \mu_2, \ldots, \mu_N) + \sum_{v=1}^{N} \frac{\partial g}{\partial \mu_v}\bigg|_{\mu} (y_v - \mu_v) + R,$$

where R contains derivatives of the second order. The first term is a constant. The second term has zero average. Under certain conditions, the effect of the R term is negligible for N large. Hence, if the Lindeberg condition (4) is satisfied for the random variables $\left\{\left.\frac{\partial g}{\partial \mu_v}\right|_\mu (y_v - \mu_v)\right\}$, we have that x is asymptotically (in N) distributed as a normal variate with mean $g(\mu_1, \mu_2, \ldots, \mu_N)$.

Many other generalizations are possible. For example, we may make N a random variable and introduce dependence between the $\{y_i\}$. Still, to assume that a random data set is normal due to some mystical ubiquity of the Central Limit Theorem would seem to be an exercise in overoptimism. Yet Galton was anything but naive. His use of the normal distribution is laced with caveats, e.g., "I am satisfied to claim that the Normal Curve is a fair average representation of the Observed Curves during nine-tenths of their course; that is, for so much of them as lies between the grades of 5% and 95%." [20, p. 57], and "It has been shown that the distribution of very different human qualities and faculties is approximately Normal, and it is inferred that with reasonable precautions we may treat them as if they were wholly so, in order to obtain approximate results" [20, p. 59].

2.2. The Johnson Family of Distributions

There is no doubting that the examination of a vast number of disparate data sets reveals that the normal distribution, for whatever reasons, is a good approximation. This being so, it is fair to hope that a nonnormal data set might be the result of passing a normal predata set through some filter. Galton suggested that one might seek to find a transformation which transformed a nonnormal data set $(x_1, x_2, \ldots, x_n)$ into a normal data set $\{z_1, z_2, \ldots, z_n\}$. One such transformation, proposed by Galton [19], was

$$(30) \qquad z = \gamma + \delta \log\left(\frac{x - \xi}{\lambda}\right).$$

Thus, if z is $N(0, 1)$, then x is said to be log normal. Moreover, since z may be written

$$z = (\gamma - \delta \log \lambda) + \delta \log(x - \xi)$$
$$= \gamma' + \delta \log(x - \xi),$$

we may estimate the three characterizing parameters from the first three sample moments. The density of $w = \dfrac{x - \xi}{\lambda}$ is given by

(31) $$g(w) = \frac{\delta}{\sqrt{2\pi}\, w} \exp\left[-\frac{1}{2}(\gamma + \delta \log w)^2 \right], \qquad w \geq 0.$$

The density is unimodal and positively skewed.

N. L. Johnson [25] (see also [27, p. 167]) has referred to data sets which may be transformed to approximate normality by the transformation in (30) as members of the "S_L system." He has suggested that one should consider a more general transformation to normality, namely,

(32) $$z = \gamma + \delta f\left(\frac{x - \xi}{\lambda}\right),$$

where f is nondecreasing in $\left(\dfrac{x - \xi}{\lambda}\right)$ and is reasonably easy to calculate. One such f is

(33) $$f(y) = \log\left(\frac{y}{1 - y}\right).$$

(Bounded) data sets which may be transformed to approximate normality by (33) are members of the "S_B system." Various methods of estimating γ and δ are given, e.g., when both endpoints are known, we have

$$\hat{\gamma} = -\bar{f}/s_f$$
$$\hat{\delta} = 1/s_f,$$

where

$$\bar{f} = \frac{1}{n} \sum_{i=1}^{n} f_i$$

$$s_f^2 = \frac{1}{n} \sum_{i=1}^{n} (f_i - \bar{f})^2$$

and

$$f_i = \log\left(\frac{x_i - \xi}{\xi + \lambda - x_i}\right), \qquad i = 1, 2, \ldots, n.$$

Densities of the S_B class are unimodal unless

$$\delta < 1/\sqrt{2}, \qquad |\gamma| < \delta^{-1}\sqrt{1 - 2\delta^2} - 2\delta \tanh^{-1}(\sqrt{1 - 2\delta^2}).$$

In the latter case, the density is bimodal. Letting $w = \dfrac{x - \xi}{\lambda}$, the characterizing density of the S_B class is

$$(34) \quad g(w) = \frac{\delta}{\sqrt{2\pi}\, w(1 - w)} \exp\left[-\frac{1}{2}\left(\gamma + \delta \log \frac{w}{1 - w} \right)^2 \right], \qquad 0 \le w \le 1.$$

A third "transformation to normality" proposed by Johnson is

$$(35) \qquad\qquad\qquad f = \sinh^{-1}(y).$$

Since

$$\mu_1' = E(x) = -\sqrt{\omega}\, \sinh(\Omega)$$

$$\mu_2 = E(x - \mu_1')^2 = \frac{1}{2}(\omega - 1)(\omega \cosh 2\Omega + 1)$$

$$\mu_3 = E(x - \mu_1')^3 = -\frac{1}{4}(\omega - 1)^2 \sqrt{\omega}\{\omega(\omega + 2)\sinh 3\Omega + 3\sinh\Omega\}$$

$$\mu_4 = E(x - \mu_1')^4 = \frac{1}{8}(\omega - 1)^2\{\omega^2(\omega^4 + 2\omega^3 + 3\omega^2 - 3)\cosh 4\Omega$$

$$+ 4\omega^2(\omega + 2)\cosh 2\Omega + 3(2\omega + 1)\}$$

where $\omega = e^{\delta^{-2}}$ and $\Omega = \gamma/\delta$,
the method of moments might be used to compute the parameters of the transformation. Data sets transformable to normality by the use of (35) are said to be members of the S_u (unbounded) system. S_u densities are unimodal with mode lying between the median and zero. They are positively or negatively skewed, according to whether γ is negative or positive [27, p. 172].
Letting $w = \dfrac{x - \xi}{\lambda}$, we have as the characterizing S_u density:

$$(36) \quad g(w) = \frac{\delta}{\sqrt{2\pi}} \frac{1}{\sqrt{w^2 + 1}} \exp\left[-\frac{1}{2}[\gamma + \delta \log\{w + \sqrt{w^2 + 1}\}]^2 \right],$$

$$-\infty \le w \le \infty.$$

The approach of Johnson is clearly philosophically related to that employed by Pearson with his differential equation family of densities. In both situations four parameters are to be estimated. The selection of a transformation by Johnson may be viewed as having a similar theoretical-heuristic mix as the selection by Pearson of his differential equation model. This selection is a solution of Fisher's problem of "specification." It is to be noted that Fisher's specification-estimation dichotomy is present in the Johnson approach. The data are used in estimating the four parameters but not in the determination of the transformation to normality.

There exists strong evidence that a number of data sets may be transformed to approximate normality by the use of one of Johnson's transformations. Moreover, we might extend the Johnson class of transformations to take care of additional sets. However, an infinite number of transformations would be required to take care of random samples from all possible continuous probability distributions. Any attempt to develop an exhaustive class of data-oriented transformations to normality would be, naturally, of the same order of complexity as the task of general probability density estimation itself.

2.3. The Symmetric Stable Distributions

We have seen that the normal distribution arises quite naturally, because a random observation may itself be the sum of random variables satisfying the conditions of Theorem 1. In situations where a set of observations is not normal, it might still be the case that each observation arises by the additivity of random variables. For example, let us suppose that an observation $x = s_n = \sum_{j=1}^{n} z_j$ where the $\{z_j\}$ are independently identically distributed with cdf F. If there exist normalizing constants, $a_n > 0$, b_n, such that the distribution of $(s_n - b_n)/a_n$ tends to some distribution G, then F is said to belong to the domain of attraction of G. It is shown by Feller [16, p. 168] that G possesses a domain of attraction if and only if G is stable. G is stable if and only if a set of i.i.d. random variables $\{w_1, w_2, \ldots, w_n\}$ with cdf G has the property that for some $0 < \alpha < 2$,

$$n^{-1/\alpha}[w_1 + w_2 + \cdots + w_n]$$

has itself cdf G.

Fama and Roll [13, 14] have considered a subset of the class of stable distributions, namely, those having characteristic function

(37) $$\varphi_x(t) = E(e^{itx}) = \exp[i\,\delta t - |ct|^\alpha],$$

where $1 \leq \alpha \leq 2$.
It is interesting to note that this family of symmetric stable distributions runs the gamut of extremely heavy tails (Cauchy, $\alpha = 1$) to sparse tails

(normal, $\alpha = 2$). It requires only three parameters to characterize a member of this family. Hence, it would appear that a methodology might be readily developed to handle the specification and estimation problem for many nonnormal data sets. There is one apparent drawback, namely, whereas the characteristic function has a very simple form, there is no known simple form for the corresponding density. Bergstrom [3] gives the series representation

$$(38) \qquad g(u) = \frac{1}{\pi\alpha} \sum_{k=0}^{\infty} (-1)^k \frac{\Gamma\left(\dfrac{2k+1}{\alpha}\right)}{(2k)!} u^{2k},$$

where $u = \dfrac{x - \delta}{c}$

and $1 < \alpha \le 2$.

Thus, Fama and Roll resorted to an empirical Monte Carlo approach for the estimation of the characterizing parameters. They suggest as an estimator for the location parameter δ the 25% trimmed mean, i.e.,

$$(39) \qquad \hat{\delta} = \frac{2}{n} \left[x_{([.25n])} + x_{([.25n]) + 1)}, \ldots, + x_{([.75n])} \right].$$

For an estimate of the scale parameter c, they suggest

$$(40) \qquad \hat{c} = \frac{1}{(.827)2} \left[x_{([.72n])} - x_{([.28n])} \right].$$

As one would expect, the most difficult parameter to be estimated is α. They suggest the estimation of α by the computation

$$\hat{v} = \frac{x_{([.97n])} - x_{([.03n])}}{2\hat{c}},$$

followed by a table look-up using a chart given in [14]. Only time will tell whether, in practice, a substantial number of nonnormal data sets can be approximated by members of the symmetric stable family. Clearly, only unimodal symmetric densities can be so approximated. Moreover, a very large data set would be required to obtain a reliable estimator for α.

In many situations most of the practically relevant information in a random sample would be obtained if we could only estimate the center of the underlying density function. One might naively assume this could easily be accomplished by using the sample mean $\bar{x} = \dfrac{1}{n} \sum_{j=1}^{n} x_j$. However, if the observations come from the Cauchy distribution with density

$$f(x) = \frac{\sigma}{\pi} \frac{1}{(x - \mu)^2 + \sigma^2}$$

and characteristic function

$$\varphi_x(t) = e^{it\mu - |t|\sigma}.$$

Then the characteristic function of $\bar{x}$ is given by

$$\varphi_{\bar{x}}(t) = \left[\varphi_x\left(\frac{t}{n}\right) \right]^n = e^{it\mu - |t|\sigma}.$$

Thus, using the sample mean from a sample of size 10^{10} would be no more effective in estimating the location parameter μ than using a single observation.

One might then resort to the use of the sample median

$$\hat{x}_{me} = x_{[n/2]}.$$

However, if the data came from a normal distribution with mean μ and variance σ^2, then

$$\frac{E[(\hat{x}_{me} - \mu)^2]}{E[(\bar{x} - \mu)^2]} \sim 1.57.$$

Thus, the efficiency of the sample median relative to that of the sample mean is only 64%.

The task of obtaining location estimators which may be used for the entire gamut of symmetrical unimodal densities and yet have high efficiency when compared to the best that could be obtained if one knew a priori the functional form of the unknown density is one of extreme complexity. We shall not attempt to cover the topic here, since the monumental work of Tukey and his colleagues on the subject has been extensively explicated and catalogued elsewhere [1] and since our study is concerned with those cases where it is important to estimate the unknown density.

However, one recent proposal [49] for estimating the location parameter μ is given by the iterative algorithm

(41)
$$\hat{\mu}_{m+1} = \frac{\displaystyle\sum_{i=1}^{n} w_{mi} x_i}{\displaystyle\sum_{i=1}^{n} w_{mi}},$$

where the *biweight function*

$$w_{mi} = (1 - u_{mi}^2)^2 \quad \text{if} \quad u_{mi}^2 < 1$$
$$= 0 \quad \text{otherwise}$$

$$u_{mi} = \frac{x_i - \hat{\mu}_m}{cs}$$

s is a scale estimator (e.g., the interquartile range or median absolute deviation)

c is an appropriately chosen constant,

and

$\hat{\mu}_0$ is the sample median.

2.4. Series Estimators

One natural approach in the estimation of a probability density function is first to devise a scheme for approximating the density, using a simple basis assuming perfect information about the true density. Then the data can be used to estimate the parameters of the approximation. Such a two-step approach is subject to some danger. We might do a fine job of devising an approximation in the presence of perfect information, which would be a catastrophe as the front end of an estimation procedure based on data.

For example, suppose we know that a density exists and is infinitely differentiable from $-\infty$ to $+\infty$. Then if we actually knew the functional form of the unknown density, we could use a Maclaurin's expansion representation

$$(42) \qquad f(x) = \sum_{j=0}^{\infty} x^j f^{(j)}(0)/j! = \sum_{j=0}^{\infty} a_j x^j.$$

But to estimate the $\{a_j\}$ from a finite data set $\{x_1, x_2, \ldots, x_n\}$ to obtain

$$(43) \qquad \hat{f}(x) = \sum_{j=0}^{\infty} \hat{a}_j x^j$$

is impossible.

If we decide to use

$$(44) \qquad \hat{f} = \sum_{j=0}^{m} \hat{\hat{a}}_j x^j \quad \text{with} \quad m \ll n$$

and are satisfied with a decent estimate over some finite interval $[a, b]$, we can come up with some sort of estimate—probably of little utility.

We might first preprocess the data via the transformation

$$(45) \qquad y_j = \frac{x_j - \text{sample median}}{\text{sample interquartile range}}.$$

This will help if the unknown density is symmetrical and unimodal. But if the density is multimodal, we are still in big trouble.

If we assume that the unknown density is close to normal, we might first transform the data via

$$y_j = \frac{x_j - \bar{x}}{s}$$

and then use an approximation which would be particularly good if y is $N(0, 1)$. Such an approximation is the Gram–Charlier series

$$(46) \qquad f(y) = \alpha(y) \sum_{j=0}^{s} a_j H_j(y)/j!$$

where $\alpha(y) = \dfrac{1}{\sqrt{2\pi}} e^{-y^2/2}$

and the Hermite polynomials $\{H_j(y)\}$ are given by

$$(47) \qquad H_j(y) = y^j - \frac{j^{[2]}y^{j-2}}{2.1!} + \frac{j^{[4]}y^{j-4}}{2^2.2!} - \frac{j^{[6]}y^{j-6}}{2^3.3!} + \cdots .$$

We recall that the Hermite polynomials are orthonormal with respect to the weight function α, i.e.,

$$(48) \qquad \int_{-\infty}^{\infty} H_i(y)H_j(y)\alpha(y)\, dy = 1 \quad \text{if} \quad i = j$$

$$= 0 \quad \text{otherwise.}$$

If we select the a_j so as to minimize

$$(49) \qquad S(a) = \int (f_0(y) - \alpha(y) \sum_{j=0}^{s} a_j H_j(y))^2 \alpha^{-1}(y)\, dy,$$

then the $\{a_j\}$ are given for all s [34, p. 26] by

$$(50) \qquad a_0 = 1$$
$$a_1 = 0$$
$$a_2 = (\mu_2 - 1)/2!$$
$$a_3 = \mu_3/3!$$
$$a_4 = (\mu_4 - 6\mu_2 + 3)/4!$$
$$a_5 = (\mu_5 - 10\mu_3)/5!$$
$$a_6 = (\mu_6 - 15\mu_4 + 45\mu_2 - 15)/6! \quad \text{etc.}$$

and

$$\mu_j = \int_{-\infty}^{\infty} y^j f_0(y) \, dy.$$

If y is $N(0, 1)$, then

$$\mu_{2k} = \frac{(2k)!}{2^k k!}$$

$$\mu_{2k+1} = 0.$$

So

$$a_j = 0 \quad \text{for} \quad j \neq 1.$$

Thus, for y $N(0, 1)$, we have simply

$$f(y) = \frac{1}{\sqrt{2\pi}} \exp\left[-\frac{y^2}{2}\right].$$

Naturally, any other density will require a more complicated Gram–Charlier representation. Nevertheless, to the extent that y is nearly $N(0, 1)$, it is reasonable to hope that the $\{a_j\}$ will quickly go to zero as j increases (in fact, this is one measure of closeness of a density to $N(0, 1)$). Cramér [9, see also 27, p. 161] has shown that if f_0 is of bounded variation in every interval, and if $\int |f_0(y)| \alpha^{-1}(y) \, dy < \infty$, then the Gram–Charlier representation (46) exists and is uniformly convergent in every finite interval.

The situation is rather grim for Gram–Charlier representations when f_0 is significantly different from $N(0, 1)$ and the coefficients $\{a_j\}$ must be estimated from data. The method of moments appears a natural means for obtaining $\{\hat{a}_j\}$. However, as a practical matter, sample moments past the fourth generally have extremely high variances. Shenton [46] has examined the quality of fitting $\hat{f}$, using the method of moments for the simple case where

(51)
$$f_0(y) = \sum_{j=0}^{4} a_j H_j(y)/j!.$$

He found that even in this idealized case, the procedure is not particularly promising. It appears clear that as a practical matter the naive Gram–Charlier estimation procedure is unsatisfactory unless f_0 is very close to α (in which case we would probably do as well to assume y is normal).

Kronmal and Tarter [29, 47, 48] have considered the representation for the density f defined on $[0, 1]$

(52)
$$f(x) = \sum_{j=-\infty}^{\infty} B_k \psi_k(x),$$

where

$$\psi_k(x) = e^{2\pi i k x}$$

and

$$B_k = \int_0^1 \psi_k(x)f(x)\,dx.$$

For the practical case, where f is unknown, they propose as an estimate of f based on the data set $\{x_1, x_2, \ldots, x_n\}$

(53)
$$\hat{f}(x) = \sum_{k=-\infty}^{\infty} \hat{\hat{B}}_k \psi_k(x),$$

where

$$\hat{B}_k = \frac{1}{n} \sum_{j=1}^{n} e^{-2\pi i k x_j}$$

and

$$\hat{\hat{B}}_k = \hat{B}_k \quad \text{if} \quad \hat{B}_k \hat{B}_{-k} > \frac{2}{n+1}, \qquad k \leq K(n)$$

$$= 0 \quad \text{otherwise.}$$

The fact that the Kronmal–Tarter approach requires that the unknown density have finite support is small liability. It would be unreasonable to expect any "nonparametric" data-based approach to give an estimate of the density outside some interval $[a, b]$ within the range of the data. A transformation from $[a, b]$ to $[0, 1]$ is easily obtained via

$$y = (x - a)/(b - a).$$

Of greater difficulty is the possibility of high frequency wiggles in the estimated density. Naturally, in order to have any hope of consistency, we should have $\lim_{n \to \infty} K(n) = \infty$. However, as a practical matter, since n is never infinite, consistency is not a particularly important criterion. Accordingly, Kronmal and Tarter suggest using $k = 10$. Of course, with such a truncation, we would not expect $\hat{f}(x)$ to be nonnegative for all x or that $\int \hat{f}(x)\,dx \equiv 1$.

More generally, let us approximate a known $f \in L^2[0, 1]$ defined on $[0, 1]$ by

(54)
$$f_N(x) = f_0(x) \sum_{k=0}^{N} a_k \varphi_k(x),$$

where

$$\int_0^1 \varphi_j(x)\varphi_k(x)f_0(x)\,dx = 1 \quad \text{if} \quad j = k$$

$$= 0 \quad \text{if} \quad j \neq k$$

$$\varphi_0(x) = 1$$

$$(55) \qquad a_k = \int_0^1 \varphi_k(x) f(x)\, dx = E_f(\varphi_k(x))$$

$f_0(x)$ is a density defined on $[0, 1]$.

Now, defining

$$(56) \qquad \|g(x)\|^2 = \int_0^1 g^2(x) f_0^{-1}(x)\, dx,$$

we have

$$(57) \quad S(a) = \left\| f - f_0(x) \sum_{k=0}^N a_k \varphi_k(x) \right\|^2$$

$$= \|f\|^2 - \sum_{k=0}^N \left(\int_0^1 f(x)\varphi_k(x)\, dx \right)^2 + \sum_{k=0}^N \left(\int_0^1 f(x)\varphi_k(x)\, dx - a_k \right)^2.$$

Thus, for any N, the $\{a_k\}$ given in (55) will minimize $S(a)$.

Now, in the practical case, where f is known only through a random sample $\{x_1, x_2, \ldots, x_n\}$, a naive estimate for f is given by

$$(58) \qquad \hat{f}_n(x) = \frac{1}{n} \sum_{j=1}^n \delta(x - x_j),$$

where the Dirac delta function $\delta(.)$ is defined via

$$\int_{-\infty}^{\infty} \delta(y)\, dy = 1$$

$$\delta(y) = 0 \quad \text{if} \quad y \neq 0.$$

Using $\hat{f}_n$ in (55), we have as an estimate for a_k

$$(59) \qquad \hat{a}_k = \int_0^1 \left(\frac{1}{n} \sum_{j=1}^n \delta(x - x_j) \right) \varphi_k(x)\, dx$$

$$= \frac{1}{n} \sum_{j=1}^n \varphi_k(x_j),$$

giving

$$(60) \qquad \hat{f}(x) = f_0(x) \sum_{k=0}^N \hat{a}_k \varphi_k(x).$$

If $\{\varphi_k\}_{k=1}^\infty$ is a basis for $L^2[0, 1]$ then (54) with $N = \infty$ gives f. Consider the integrated mean square error of the estimator in (60)

$$(61) \qquad \text{MISE} = E_f \int_0^1 (f(x) - \hat{f}(x))^2 f_0(x)\, dx$$

$$= \sum_{k=N+1}^\infty a_k^2 + \frac{1}{n} \left[\sum_{k=0}^N \int_0^1 \varphi_k^2(x) f(x)\, dx - \sum_{k=0}^N a_k^2 \right].$$

We note by the Cauchy–Schwarz Inequality that the second term in (61) is nonnegative. The first term in (61) shows that for the MISE to go to zero, it is necessary that N go to infinity, as n goes to infinity. However, the second term shows that N must not go to infinity too rapidly relative to the sample size n.

Watson [54] considers the estimate

$$\hat{f}_w(x) = \sum_{k=0}^{\infty} \lambda_k(n) \hat{a}_k \varphi_k(x),$$

where the $\{\lambda_k(n)\}$ are chosen so as to minimize the mean integrated square error

$$(62) \qquad J(\lambda(n)) = E_f \int (f(x) - \hat{f}_w(x))^2 \, dx$$

$$= \sum_{k=0}^{\infty} \left\{ a_k^2 (1 - \lambda_k(n))^2 + n^{-1} \lambda_k^2(n) \operatorname*{var}_f (\varphi_k(x)) \right\}.$$

Hence, $\lambda_k(n) = a_k^2 \Big/ \left[a_k^2 + n^{-1} \operatorname*{var}_f (\varphi_k(x)) \right]$.

Unfortunately, we know neither the $\{a_k\}$ nor $\left\{ \operatorname*{var}_f (\varphi_k(x)) \right\}$.

One ad hoc approach, not suggested by Watson, would be to replace a_k and $(\operatorname{var}(\varphi_k))$ in the expression for $\lambda_k(n)$ by their representations, using a prior guess f_0 for the unknown density:

$$\lambda_k(n) = \begin{cases} \left[1 + n^{-1} \dfrac{1 + \int \varphi_k^2(x) f_0(x) \, dx}{\left(\int \varphi_k(x) f_0(x) \, dx \right)^2} \right]^{-1} & \text{for } k \le \sqrt{n} \\[4ex] 0 & \text{for } k > \sqrt{n} \end{cases}$$

Brunk [7] uses a Bayesian approach to resolve this difficulty. He notes that it is quite natural to use as f_0 our best prior guess for f. Then natural prior guesses for the a_k are

$$(63) \qquad \begin{aligned} E[a_k] &= 1 \quad \text{for } k = 0 \\ &= 0 \quad \text{for } k > 0. \end{aligned}$$

The second prior moments are more difficult to postulate. Brunk suggests

$$(64) \qquad \begin{aligned} \operatorname{Cov}(a_j, a_k) &= \sigma_j^2 \quad \text{if } j = k \\ &= 0 \quad \text{otherwise.} \end{aligned}$$

Then, to minimize

(65)
$$S(\lambda(n)) = E\left\{E_f \sum_{k=0}^{\infty} [\lambda_k(n)\hat{a}_k - a_k]^2 | a\right\}$$

we have

(66)
$$\lambda_k(n) = \frac{n\sigma_k^2}{(n-1)\sigma_k^2 + E[E_f(\varphi_k^2(x)|a]}.$$

But if $f \sim f_0$, then it is reasonable to replace E_f by E_{f_0} to give

(67)
$$E[E_{f_0}(\varphi_k^2(x))|a] = 1.$$

Thus, we have

(68)
$$\lambda_0(n) = 1$$

$$\lambda_k(n) = \frac{n}{n - 1 + (1/\sigma_k^2)} \quad \text{for} \quad k > 0.$$

Using the trigonometric basis

(69)
$$\varphi_0(x) = 1 \qquad 0 \le x \le 1$$
$$\varphi_{2k-1}(x) = \sqrt{2}\,\sin(2\pi kx)$$
$$\varphi_{2k}(x) = \sqrt{2}\,\cos(2\pi kx), \qquad k = 1, 2, \ldots$$
$$f_0(x) = 1, \qquad 0 \le x \le 1,$$
$$E[E_f(\varphi_k^2(x)|a)] = 1 \quad \text{for all} \quad k.$$

Brunk gives several empirical alternatives for estimating $\lambda_k(n)$ when this basis is used. For example, let

(70)
$$\xi_k = \left[\frac{1}{2}\left\{\hat{a}_{2k-1}^2 + \hat{a}_{2k}^2\right\} - \frac{1}{n}\right].$$

Then, if

(71)
$$\gamma_k = \sigma_{2k-1}^2 = \sigma_{2k}^2, \qquad k = 1, 2, \ldots$$

let

(72)
$$\hat{\gamma}_k = (1/4t_k^2)t_k + [1/n + (n-1)t_k'/n]^{-2}t_k',$$

where

$$t_k = (1/5)^{k-1}$$
$$t_k' = \max(\xi_k, 0), \qquad k = 1, 2, \ldots .$$

Using the resulting estimates in (66), we obtain a sequence of damping factors $\{\lambda_k(n)\}$. Brunk notes that many procedures for choosing a sequence of γ_k, which decrease smoothly to 0 as k increases, appear to work better than any rule of the form

$$(73) \qquad \lambda_k(n) = 1 \quad \text{for} \quad k \leq M$$
$$= 0 \quad \text{for} \quad k > M.$$

Let us now consider the following iterative algorithm for estimating an unknown density f with domain of positivity $[a, b]$. We assume a random sample $\{x_1, x_2, \ldots, x_n\}$ is given. Also we shall assume we have a prior guess as to the density—say $\hat{f}_0$ (the parameters characterizing $\hat{f}_0$ may be estimated by classical means, e.g., maximum likelihood). Let $\{\varphi_j^{(1)}\}$ be an appropriate orthonormal (with respect to $\hat{f}_0$) basis. Then we take as our estimate for f

$$(74) \qquad \hat{f}_1(x) = \hat{f}_0(x) \sum_{j=0}^{N(1)} \hat{a}_j^{(1)} \varphi_j^{(1)}(x),$$

where

$$\hat{a}_0^{(1)} = 1$$

$$\hat{a}_j^{(1)} = \frac{1}{n} \sum_{i=1}^{n} \varphi_j^{(1)}(x_i), \qquad j > 1.$$

Using $\hat{f}_1$ as a new weight function, we obtain a new orthonormal basis $\{\varphi_j^{(2)}\}$. Then our second estimate for f is

$$(75) \qquad \hat{f}_2(x) = \hat{f}_1(x) \sum_{j=0}^{N(2)} a_j^{(2)} \varphi_j^{(2)}(x),$$

where

$$\hat{a}_0^{(2)} = 1$$

$$\hat{a}_j^{(2)} = \frac{1}{n} \sum_{i=1}^{n} \varphi_j^{(2)}(x_i), \qquad j > 1.$$

If the algorithm converges to a result close to the true f, it is reasonable to hope that after a few iterations $N(m)$ can be taken to be small. Naturally, a crucial consideration here is the selection of an appropriate basis $\{\varphi_j^{(m)}\}$ at each stage of the process. We as yet have no good answers as to how this task may be accomplished and merely state it as an open problem.

As a procedure related to series estimation, we consider the following approach based on cubic B-splines. We shall assume that the data has been transformed to lie in $[-M(n), M(n)]$ where n is the size of a random sample $\{x_1, x_2, \ldots, x_n\}$ and $M(n)$ is an integer. Let

$$(76) \qquad b_k(x) = \frac{1}{6} \sum_{j=-2}^{2} (-1)^{j+2} \binom{4}{j+2} (k+j-x)_+^3,$$

where

$$(y)_+ = y \quad \text{if} \quad y > 0$$
$$= 0 \quad \text{otherwise.}$$

$b_k(x)$ is readily seen to be a probability density symmetric about k. Our estimator for the unknown density f is

$$(77) \qquad f_b(x) = \sum_{k=-M(n)+2}^{M(n)-2} c_k b_k(x),$$

where $(c_{-M(n)+2}, c_{-M(n)+1}, \ldots, c_{+M(n)-2})$ is chosen to maximize $\displaystyle\prod_{i=1}^{n} f_b(x_i)$ subject to the constants

$$\sum c_k = 1, c_k \geq 0 \;\forall k.$$

Clearly, $M(n)$ should go to infinity as $n \to \infty$, but at a rate slower than n. Perhaps $M(n) = 0(\sqrt{n})$ is a practical value. Again, we have not investigated this matter and state it as an open problem.

2.5. Kernel Estimators

Of the methods used to estimate probability densities of unknown functional form, the most used is the histogram. Let us suppose we have a random sample $\{x_1, x_2, \ldots, x_n\}$ from an unknown absolutely continuous probability density with domain of positivity $[a, b]$. In the event that the unknown density, say $g(x)$, has infinite range, we shall content ourselves with estimating the truncated density

$$(78) \qquad f(x) = \frac{g(x)}{\int_a^b g(t)\, dt} \quad \text{for} \quad a \leq x \leq b$$
$$= 0 \quad \text{otherwise.}$$

In the following, we shall assume that those points outside $[a, b]$ have been discarded and that each of the $\{x_i\}_{i=1}^n$ is in the interval $[a, b]$. As a practical matter, if the density g has domain $[-\infty, \infty]$, then as the sample size increases we will take a to be smaller and smaller, and b to be larger and larger.

Let us partition $[a, b]$ by $a = t_0 < t_1 < \cdots < t_m = b$. Then we shall obtain an estimator f_H of the form

(79)
$$f_H(t) = c_i \quad \text{for} \quad t_i \leq t < t_{i+1}, \qquad i = 0, \ldots, m-1$$
$$f_H(b) = c_{m-1}$$
$$f_H(t) = 0 \quad \text{for} \quad t \notin [a, b],$$

where $f_H(t) \geq 0$ and $\int_a^b f_H(t) \, dt = 1$.

If q_i is the number of observations falling in the ith interval, then for $\hat{f}_H$ we shall use

(80)
$$\hat{c}_i = \frac{q_i}{n(t_{i+1} - t_i)} \quad \text{for} \quad i = 0, \ldots, m-1.$$

The intuitive appeal of $\hat{f}_H$ is clear. The number of observations falling into each of the intervals is a multinomial variate. Thus, q_i/n estimates $\int_{t_i}^{t_{i+1}} f(t) \, dt$. Since we have assumed that f is absolutely continuous, if $t_{i+1} - t_i$ is small, then $f(t) \sim f(t_i)$ for $t_i \leq t < t_{i+1}$. Hence, q_i/n estimates $(t_{i+1} - t_i)f(t)$, and

$$\frac{q_i}{n(t_{i+1} - t)}$$ estimates $f(t)$.

Theorem 2. Among estimates of the form (79) $\hat{f}_H$ uniquely maximizes the likelihood

(81)
$$L(c_0, \ldots, c_{m-1}) = \prod_{j=1}^n f_H(x_j)$$
$$= \prod_{i=0}^{m-1} c_i^{q_i}$$

Proof. The feasible set $C \subset R^m$ must satisfy

(82)
$$\sum_{i=0}^{m-1} c_i(t_{i+1} - t_i) = 1$$

and

$$c_i \geq 0.$$

But L is a continuous function of c and C is compact. Thus there exists a global maximizer c^*. Let us assume, without loss of generality, that each interval contains at least one observation, since if this were not the case, say for the ith interval, a moments reflection should convince the reader that c_i^* must be equal to zero. Moreover, if $c_i^* = 0$ for any i, then $L(c^*) = 0$. But $c' = \dfrac{1}{m}\left(\dfrac{1}{t_1 - t_0}, \dfrac{1}{t_2 - t_1}, \ldots, \dfrac{1}{t_m - t_{m-1}}\right) \in C$ and $L(c') > 0$. Hence, the inequality constraints are not active at the solution and from Appendix I.5 there exists $\lambda \in R$, such that

$$(83) \qquad \nabla L(c^*) = \lambda(t_1 - t_0, t_2 - t_1, \ldots, t_m - t_{m-1}).$$

From (81), (82), and (83), we have

$$(84) \qquad q_i L(c^*) = \lambda c_i^*(t_{i+1} - t_i)$$

$$\sum_{i=0}^{m-1} q_i L(c^*) = \lambda$$

$$nL(c^*) = \lambda.$$

Substituting for λ in (84), we have

$$q_i L(c^*) = nL(c^*)c_i^*(t_{i+1} - t_i).$$

So

$$(85) \qquad c_i^* = \frac{q_i}{n(t_{i+1} - t_i)}.$$

But any maximizer must satisfy (85), and we have earlier shown that a maximizer exists. ∎

Theorem 3. Suppose that f has continuous derivatives up to order three except at the endpoints of $[a, b]$, and f is bounded on $[a, b]$. Let the mesh be equal spacing throughout $[a, b]$, so that $t_{i+1,n} - t_{i,n} = 2h_n$. Then if $n \to \infty$ and $h_n \to 0$, so that $nh_n \to \infty$, then for $x \in (a, b)$

$$\mathrm{MSE}(\hat{f}_H(x)) = E[(\hat{f}_H(x) - f(x))^2] \to 0,$$

i.e., $\hat{f}_H(x)$ is a consistent estimator for $f(x)$.

Proof. For any h_n, we adopt the rule that $[a, b]$ is to be divided into $k = [(b - a)/(2h_n)]$ interior intervals, with two intervals of length $(b - a - 2kh_n)/2$ at the ends. Then for any h_n, x is unambiguously contained in an interval with a well-defined midpoint $x' = x'(x, h_n)$.

Expanding f about x', we have

(86) $\quad f(x) = f(x') + f'(x')(x - x') + \frac{1}{2}f''(x')(x - x')^2 + 0(x - x')^3,$

where

$\quad\quad x \in [t_k, t_{k+1}), \quad \text{and} \quad t_{k+1} - t_k = 2h_n \quad \text{for} \quad k = 0, 1, \ldots, m - 1.$

Letting $p_k = \int_{t_k}^{t_{k+1}} f(x)\, dx$,
we have

$$p_k = 2h_n f(x') + \frac{h_n^3}{3}f''(x') + 0(h_n^4).$$

Now,

(87) $\quad\quad E[(\hat{f}_H(x') - f(x'))^2] = \text{Var}(\hat{f}_H(x')) + \text{Bias}^2(\hat{f}_H(x')),$

where

$$\text{Var}(\hat{f}_H(x')) = E\left(\frac{q_k}{2nh_n} - \frac{p_k}{2h_n}\right)^2 = \frac{p_k(1 - p_k)}{4nh_n^2}$$

$$= \frac{1}{4nh_n}\left[2f(x') - 4f^2(x')h_n + \frac{h_n^2}{3}f''(x') + 0(h_n^3)\right]$$

$$= \frac{1}{2nh_n}\left[f(x') - 2h_n f^2(x') + \frac{h_n^2}{6}f''(x') + 0(h_n^3)\right]$$

and

$$\text{Bias}^2(\hat{f}_H(x')) = \left[\frac{p_k}{2h_n} - f(x')\right]^2$$

$$= \frac{h_n^4}{36}(f''(x'))^2 + 0(h_n^5).$$

Thus,

(88) $\quad\quad \text{MSE}(\hat{f}_H(x')) = \frac{f(x')}{2nh_n} + \frac{h_n^4}{36}(f''(x'))^2 + 0\left(\frac{1}{n}\right) + 0(h_n^5).$

Thus, if $nh_n \to \infty$, $n \to \infty$, $h_n \to 0$,

$$\text{MSE}(\hat{f}_H(x')) \to 0.$$

Now, let x be any point in $[t_k, t_{k+1})$. We recall that $\hat{f}_H(x) = \hat{f}_H(x')$. Then

$$\text{MSE}(\hat{f}_H(x)) = E[\hat{f}_H(x) - f(x))^2] = E[(\hat{f}_H(x') - f(x') + f(x') - f(x))^2]$$
$$\leq 2E[(\hat{f}_H(x') - f(x'))^2] + 2[(f(x') - f(x))^2].$$

Assuming the spacing is sufficiently fine that $|f(x') - f(x)|$ is no greater than $|f(x') - f(t_k)|$ or $|f(x') - f(t_{k+1})|$, we have

(89) $\text{MSE}(\hat{f}_H(x)) \leq \dfrac{f(x')}{nh_n} + 2|f'(x')|^2 h_n^2 + 0\left(\dfrac{1}{n}\right) + 0(h_n^3).$

Thus if $nh_n \to \infty$, $n \to \infty$, $h_n \to 0$, $\text{MSE}(\hat{f}_H(x)) \to 0$. ∎

Remark. It is interesting to note from (88) and (89) that if we select

$$h_n = \left[\frac{9f(x')}{2(f''(x'))^2}\right]^{1/5} n^{-1/5},$$

then the mean square error at x' is of order $n^{-4/5}$. However, this choice of $\{h_n\}$ gives a MSE away from x' as high as order $n^{-2/5}$. On the other hand, we may, viz. (89) select $h_n = \left[\dfrac{f(x')}{4(f'(x'))^2}\right]^{1/3} n^{-1/3}$ to obtain convergence throughout the kth interval of order $n^{-2/3}$.

Integrating (89) over $[a, b]$, we may minimize our bound on the integrated mean square error by selecting

(90) $$h_n = \left[\frac{1}{4 \int (f'(x))^2 \, dx}\right]^{1/3} n^{-1/3}$$

to obtain

(91) $\displaystyle \int \text{MSE}(\hat{f}_H(x)) = \text{IMSE} \leq 3\left[\frac{1}{2}\int (f')^2 \, dx\right]^{1/3} n^{-2/3} + 0\left(\frac{1}{n} + h_n^3\right).$

Now the likelihood in (81) is, of course, the joint density of $(x_1, x_2, \ldots, x_n)$ given $c = (c_0, c_1, \ldots, c_{m-1})$; i.e.,

(92) $$L(c) = L(x_1, x_2, \ldots, x_n|c).$$

Let us consider as a prior distribution for $c' = (c_0(t_1 - t_0), \ldots, c_{m-1}(t_m - t_{m-1}))$

(93) $P(c_0', c_1', \ldots, c_{m-1}') = \dfrac{\Gamma(\theta_0 + \cdots + \theta_{m-1})}{\Gamma(\theta_0) \cdots \Gamma(\theta_{m-1})} c_0'^{\theta_0 - 1} \cdots c_{m-1}'^{\theta_{m-1} - 1}$

$$\text{if } \sum_{i=0}^{m-1} c_i' = 1, \quad c_i' > 0, \quad i = 0, \ldots, m-1$$

$$= 0 \quad \text{otherwise.}$$

Then the joint density of (x, c) is given by

(94) $\displaystyle s(x_1, \ldots, x_n, c_0', \ldots, c_{m-1}') = \prod_{i=0}^{m-1} c_i'^{q_i + \theta_i - 1} \frac{\Gamma(\theta_0 + \cdots + \theta_{m-1})}{\Gamma(\theta_0) \cdots \Gamma(\theta_{m-1})} h_n^{-n}.$

The marginal density of x, then, is given by

$$(95) \quad u(x_1, \ldots, x_n) = \left[\prod_{i=0}^{m-1} \frac{\Gamma(\theta_i + q_i)}{\Gamma(\theta_i)}\right] \frac{\Gamma(\theta_0 + \cdots + \theta_{m-1})}{\Gamma(\theta_0 + \cdots + \theta_{m-1} + n)} h_n^{-n}.$$

The posterior density of c' is then given by

$$(96) \quad v(c_0', \ldots, c_{m-1}' | x) = \frac{s(x_1, \ldots, x_n, c_0', \ldots, c_{m-1}')}{u(x_1, \ldots, x_n)}$$

$$= \frac{\Gamma(\theta_0 + \cdots + \theta_{m-1} + n)}{\Gamma(\theta_0 + q_0) \cdots \Gamma(\theta_{m-1} + q_{m-1})} \prod_{i=0}^{m-1} c_i'^{q_i + \theta_i - 1}.$$

Now, the Dirichlet distribution with parameter $(\theta_0, \ldots, \theta_{m-1})$ has

$$(97) \qquad\qquad E(c_i') = \frac{\theta_i}{\sum_j \theta_j}$$

$$(98) \qquad\qquad \mathrm{Var}(c_i') = \frac{\theta_i(\theta_0 + \cdots + \theta_{m-1} - \theta_i)}{(\sum \theta_j)^2(\sum \theta_j + 1)}$$

$$(99) \qquad\qquad \mathrm{Cov}(c_i', c_j') = \frac{-\theta_i \theta_j}{(\sum \theta_k)^2(\sum \theta_k + 1)}.$$

So the posterior mean of c' is given by

$$(100) \qquad\qquad c_i' = \frac{\theta_i + q_i}{\sum_{j=0}^{m-1} \theta_j + n},$$

giving

$$(101) \qquad c_i = \frac{\theta_i + q_i}{\left(\sum_{j=0}^{m-1} \theta_j + n\right)(t_{i+1} - t_i)}, \qquad i = 0, \ldots, m-1.$$

Now a ready guess is available for the prior parameter $(\theta_0, \ldots, \theta_{m-1})$ via (97). We may replace $E(c_i') = \frac{\theta_i}{\sum \theta_j}$ by our prior feelings as to the number of observations which will fall into the ith interval, i.e., let

$$(102) \qquad\qquad \frac{\theta_i}{\sum \theta_j} = \int_{t_i}^{t_{i+1}} f_p(x) \, dx,$$

where f_p is our best guess as to the true f. This will give us $m - 1$ equations for $\{\theta_0, \theta_1, \ldots, \theta_{m-1}\}$. The additional equation may be obtained from our feelings as to the variance of c_i for some i. Clearly, for fixed spacing and $(\theta_0, \ldots, \theta_{m-1})$, as $n \to \infty$, the posterior mean of c is the maximum likelihood estimator given in (85). Note, moreover, that if we wish to use the posterior mean estimator in such a way that we obtain a consistent estimator f, we may easily do so by simply requiring that $\underset{i}{\text{Min}} \, (t_{i+1} - t_i)n \to \infty$ and $\underset{i}{\text{Max}} \, (t_{i+1} - t_i) \to 0$ as $n \to \infty$. The method for choosing prior values for $(\theta_0, \ldots, \theta_{m-1})$ for m of arbitrary size is that suggested in the discussion following (101).

We note, then, that the consistent histogram is a nonparametric estimator, in the sense that the number of characterizing parameters increases without limit as the sample size goes to infinity. It has the advantage over the classical series estimation of being more local. However, it is discontinuous, completely ad hoc, harder to update, and typically requires more parameters for a given sample size than the orthonormal series estimators. Clearly, the histogram should offer interesting possibilities for generalization.

In 1956 Murray Rosenblatt, in a very insightful paper [38], extended the histogram estimator of a probability density. Based on a random sample $\{x_i\}_{i=1}^n$ from a continuous but unknown density f, the Rosenblatt estimator is given by

$$(103) \qquad \hat{f}_n(x) = \frac{\#\text{ sample points in } (x - h_n, x + h_n)}{2nh_n},$$

where h_n is a real valued number constant for each n, i.e.,

$$(104) \qquad \hat{f}_n(x) = \frac{F_n(x + h_n) - F_n(x - h_n)}{2h_n},$$

where

$$F_n(x) = \frac{\#\text{ sample points} \leq x}{n}.$$

To obtain the formula for the mean square error of $\hat{f}_n$, Rosenblatt notes that if we partition the real line into three intervals $\{x | x \leq x_1\}$, $\{x | x_1 < x \leq x_2\}$ and $\{x | x > x_2\}$, and let $Y_1 = F_n(x_1)$, $Y_2 = F_n(x_2) - F_n(x_1)$ and $Y_3 = 1 - F_n(x_2)$, then (nY_1, nY_2, nY_3) is a trinomial random variable with probabilities $(F(x_1), F(x_2) - F(x_1), 1 - F(x_2)) = (p_1, p_2, p_3)$, where F is the cumulative distribution function of x, i.e., $F(x) = \int_{-\infty}^{x} f(t)\, dt$. Thus, we have

$$(105) \qquad E[F_n(x)] = F(x)$$

(106) $\mathrm{Cov}[F_n(x_1), F_n(x_2)]$

$$= \frac{1}{n^2} E[(nY_1 - nF(x_1))(n(Y_1 + Y_2) - n(F(x_2) + F(x_1)))]$$

$$= \frac{1}{n^2} \mathrm{Cov}(nY_1, nY_2) + \frac{1}{n^2} \mathrm{Var}(nY_1)$$

$$= \frac{-1}{n^2} nF(x_1)(F(x_2) - F(x_1)) + \frac{1}{n^2} nF(x_1)(1 - F(x_1))$$

$$= -\frac{F(x_1)F(x_2)}{n} + \frac{F(x_1)}{n},$$

assuming without loss of generality that $x_1 < x_2$.

Thus, making no restrictions on x_1 and x_2,

(107)

$\mathrm{Cov}[\hat{f}_n(x_1), \hat{f}_n(x_2)]$

$$= E\left[\left(\frac{F_n(x_1 + h_n) - F_n(x_1 - h_n)}{2h_n} - \frac{F(x_1 + h_n) - F(x_1 - h_n)}{2h_n}\right)\right.$$

$$\left. \times \left(\frac{F_n(x_2 + h_n) - F_n(x_2 - h_n)}{2h_n} - \frac{F(x_2 + h_n) - F(x_2 - h_n)}{2h_n}\right)\right]$$

$$= \frac{1}{4h_n^2} \{\mathrm{Cov}[F_n(x_1 + h_n), F_n(x_2 + h_n)] - \mathrm{Cov}[F_n(x_1 + h_n), F_n(x_2 - h_n)]$$

$$- \mathrm{Cov}[F_n(x_1 - h_n), F_n(x_2 + h_n)] + \mathrm{Cov}[F_n(x_1 - h_n), F_n(x_2 - h_n)]\}$$

$$= \frac{1}{4h_n^2 n} [-F(x_1 + h_n)F(x_2 + h_n) + F(\min(x_1 + h_n, x_2 + h_n))$$

$$+ F(x_1 + h_n)F(x_2 - h_n) - F(\min(x_1 + h_n, x_2 - h_n))$$
$$+ F(x_1 - h_n)F(x_2 + h_n) - F(\min(x_1 - h_n, x_2 + h_n))$$
$$- F(x_1 - h_n)F(x_2 - h_n) + F(\min(x_1 - h_n, x_2 - h_n))].$$

If $x_1 = x_2 = x$, then

(108) $\mathrm{Var}[\hat{f}_n(x)] = \dfrac{1}{4h_n^2 n} [-F^2(x + h_n) + F(x + h_n) + F(x - h_n)F(x + h_n)$

$$- F(x - h_n) + F(x - h_n)F(x + h_n) - F(x - h_n)$$
$$- F^2(x - h_n) + F(x - h_n)]$$

$$= \frac{1}{4h_n^2 n} [F(x + h_n) - F(x - h_n) - (F(x + h_n) - F(x - h_n))^2].$$

Then, if we pick $x_1 < x_2$ and h_n sufficiently small that $x_1 + h_n < x_2 - h_n$

(109)

$\mathrm{Cov}[\hat{f}_n(x_1), \hat{f}_n(x_2)]$

$$= E\left[\left(\frac{F_n(x_1 + h_n) - F_n(x_1 - h_n)}{2h_n} - \frac{F(x_1 + h_n) - F(x_1 - h_n)}{2h_n}\right)\right.$$
$$\left. \times \left(\frac{F_n(x_2 + h_n) - F_n(x_2 - h_n)}{2h_n} - \frac{F(x_2 + h_n) - F(x_2 - h_n)}{2h_n}\right)\right]$$

$$= \frac{1}{4h_n^2}[\mathrm{Cov}(F_n(x_1 + h_n), F_n(x_2 + h_n)) - \mathrm{Cov}(F_n(x_1 + h_n), F_n(x_2 - h_n))$$
$$- \mathrm{Cov}(F_n(x_1 - h_n), F_n(x_2 + h_n)) + \mathrm{Cov}(F_n(x_1 - h_n), F_n(x_2 - h_n))]$$

$$= \frac{1}{4h_n^2}\left[-\frac{F(x_1 + h_n)F(x_2 + h_n)}{n} + \frac{F(x_1 + h_n)}{n} + \frac{F(x_1 + h_n)F(x_2 - h_n)}{n}\right.$$
$$- \frac{F(x_1 + h_n)}{n} + \frac{F(x_1 - h_n)F(x_2 + h_n)}{n} - \frac{F(x_1 - h_n)}{n}$$
$$\left. - \frac{F(x_1 - h_n)F(x_2 - h_n)}{n} + \frac{F(x_1 - h_n)}{n}\right]$$

$$= \frac{1}{4h_n^2 n}[-F(x_1 + h_n)[F(x_2 + h_n) - F(x_2 - h_n)]$$
$$+ F(x_1 - h_n)[F(x_2 + h_n) - F(x_2 - h_n)]]$$

$$= \frac{-1}{4h_n^2 n}[(F(x_2 + h_n) - F(x_2 - h_n))(F(x_1 + h_n) - F(x_1 - h_n))]$$

$$= \frac{-1}{n}\frac{F(x_2 + h_n) - F(x_2 - h_n)}{2h_n}\frac{F(x_1 + h_n) - F(x_2 - h_n)}{2h_n}$$

$$= \frac{-1}{n}f(x_1)f(x_2) - \frac{h_n^2}{6n}[f(x_1)f''(x_2) + f(x_2)f''(x_1)] + 0\left(\frac{h_n^3}{n}\right)$$

(assuming f is thrice differentiable at x_1 and x_2).
Now,

(110) $\mathrm{MSE}(\hat{f}_n(x)) = E[(\hat{f}_n(x) - f(x))^2] = \mathrm{Var}[\hat{f}_n(x)] + \mathrm{Bias}^2[\hat{f}_n(x)]$

$$= \frac{1}{4h_n^2 n}[F(x + h_n) - F(x - h_n) - (F(x + h_n) - F(x - h_n))^2]$$

$$+ \left[\frac{1}{2h_n}(F(x + h_n) - F(x - h_n)) - f(x)\right]^2.$$

But $F(x + h_n) - F(x - h_n) = 2h_n f(x) + \dfrac{h_n^3}{3} f''(x) + 0(h_n^4)$, (assuming f is thrice differentiable at x).

(111) $\text{MSE}(\hat{f}_n(x)) = \dfrac{1}{4h_n^2 n}\left[2h_n f(x) + \dfrac{h_n^3}{3} f''(x) + 0(h_n^4) \right.$

$\left. - \left(2h_n f(x) + \dfrac{h_n^3}{3} f''(x) + 0(h_n^4) \right)^2 \right]$

$+ \dfrac{1}{4h_n^2}\left[2h_n f(x) + \dfrac{h_n^3}{3} f''(x) + 0(h_n^4) - 2h_n f(x) \right]^2$

$= \dfrac{f(x)}{2h_n n} + \dfrac{h_n^4}{36}(f''(x))^2 + 0\left(\dfrac{1}{h_n n} + h_n^4 \right).$

So, if $h_n \to 0$ as $n \to \infty$ in such a way that $nh_n \to \infty$, $\text{MSE}(\hat{f}_n(x)) \to 0$. Holding f, x, and n constant; we may minimize the first two terms in (111) using a straightforward argument, yielding

(112) $$\hat{h}_n = \left[\dfrac{9}{2} \dfrac{f(x)}{(f''(x))^2} \right]^{1/5} n^{-1/5},$$

with corresponding asymptotic (in n) mean square error of

(113) $$\text{MSE}(\hat{f}_n(x)) = \dfrac{5}{4} 9^{-1/5} 2^{-4/5} |f(x)|^{4/5}(|f''(x)|)^{2/5} n^{-4/5}.$$

Studies of a variety of measures of consistency of density estimates are given in [4, 28, 33, 42, 51, 61].

The comparison between the Rosenblatt kernel estimator and the histogram estimator is of some interest. In essence, Rosenblatt's approach is simply a histogram which, for estimating the density of x, say, has been shifted so that x lies at the center of a mesh interval. For evaluating the density at another point, say y, the mesh is shifted again, so that y is at the center of a mesh interval. The fact that the MSE of the Rosenblatt procedure decreases like $n^{-4/5}$ instead of as slowly as $n^{-2/3}$, as with the fixed grid histogram, demonstrates the advantage of having the point at which f is to be estimated at the center of an interval and is reminiscent of the advantages of central differences in numerical analysis. We note that the advantage of the "shiftable" histogram approach lies in a reduction of the bias of the resulting estimator.

The shifted histogram estimator has another representation, namely,

(114) $$\hat{f}_n(x) = \dfrac{1}{n} \sum_{j=1}^{n} \dfrac{1}{h_n} w\left(\dfrac{x - x_j}{h_n} \right),$$

where $w(u) = \frac{1}{2}$ if $|u| < 1$
        $= 0$ otherwise,
and the $\{x_j\}_{j=1}^n$ are the data points.
Thus, it is clear that for all $\{x_j\}_{j=1}^n$

$$\int \hat{f}_n(x) \, dx = 1$$

and

$$\hat{f}_n(x) \geq 0.$$

A shifted histogram is, like the fixed grid histogram, always a bona-fide probability density function.

To pick a global value for h_n, we may examine the integrated mean square error

$$\int \text{MSE}(\hat{f}_n(x)) \, dx = \text{IMSE} \sim \frac{1}{2h_n n} + \frac{h_n^4}{36} \int_{-\infty}^{\infty} (f''(x))^2 \, dx + o\left(\frac{1}{h_n n} + h_n^4\right)$$

to give

(115)
$$h_n = \left[\frac{9}{2 \int (f''(x))^2 \, dx}\right]^{1/5} n^{-1/5},$$

and thus

(116)
$$\text{IMSE} \sim 2^{-4/5} 9^{-1/5} \frac{5}{4} \left[\int (f''(x))^2 \, dx\right]^{1/5} n^{-4/5}.$$

Although Rosenblatt suggested generalizing (114) to estimators using different bases than step functions, the detailed explication of kernel estimators is due to Parzen [35]. We shall consider as an estimator for $f(x)$

(117)
$$\hat{f}_n(x) = \int_{-\infty}^{\infty} \frac{1}{h_n} K\left(\frac{x-y}{h_n}\right) dF_n(y) = \frac{1}{nh_n} \sum_{j=1}^{n} K\left(\frac{x-x_j}{h_n}\right),$$

where

(118)
$$\begin{cases} \int_{-\infty}^{\infty} |K(y)| \, dy < \infty \\ \underset{-\infty < y < \infty}{\text{Sup}} \ |K(y)| < \infty \\ \lim_{y \to \infty} |yK(y)| = 0 \end{cases}$$

and

(119)
$$K(y) \geq 0$$

$$\int_{-\infty}^{\infty} K(y) \, dy = 1.$$

It is interesting to note that, if $k(\cdot)$ is an even function,

$$\hat{\mu} = \int x \hat{f}_n(x)\, dx = \frac{1}{n} \sum_{j=1}^{n} \int \frac{x}{h_n} K\left(\frac{x - x_j}{h_n}\right) dx$$

$$= \frac{1}{n} \sum_{j=1}^{n} x_j = \bar{x}$$

and

$$\hat{\sigma}^2 = \int (x - \hat{\mu})^2 \hat{f}_n(x)\, dx$$

$$= \frac{1}{n} \sum_{j=1}^{n} (x_j - \bar{x})^2 + h_n^2 \int x^2 K(x)\, dx$$

$$= s^2 + h_n^2 \int x^2 K(x)\, dx.$$

Parzen's consistency argument is based on the following lemma of Bochner [2].

Lemma. Let K be a Borel function satisfying (118). Let $g \in L^1$ (i.e., $\int |g(y)|\, dy < \infty$). Let

(120)
$$g_n(x) = \frac{1}{h_n} \int K\left(\frac{y}{h_n}\right) g(x - y)\, dy,$$

where $\{h_n\}$ is a sequence of positive constants having $\lim_{n \to \infty} h_n = 0$. Then if x is a point of continuity of g,

(121)
$$\lim_{n \to \infty} g_n(x) = g(x) \int_{-\infty}^{\infty} K(y)\, dy.$$

Proof.

$$\left| g_n(x) - g(x) \int_{-\infty}^{\infty} K(y)\, dy \right|$$

$$= \left| \int_{-\infty}^{\infty} \{g(x - y) - g(x)\} \frac{1}{h_n} K\left(\frac{y}{h_n}\right) dy \right|$$

$$\leq \sup_{|y| \leq \delta} |g(x - y) - g(x)| \int |K(z)|\, dz$$

$$+ \int_{|y| \geq \delta} \frac{|g(x - y)|}{y} \frac{y}{h_n} K\left(\frac{y}{h_n}\right) dy + |g(x)| \int_{|y| \geq \delta} \frac{1}{h_n} K\left(\frac{y}{h_n}\right) dy$$

$$\leq \sup_{|y| \leq \delta} |g(x - y) - g(x)| \int_{-\infty}^{\infty} |K(z)|\, dz$$

$$+ \frac{1}{\delta} \sup_{|z| \geq \delta/h_n} |z K(z)| \int_{-\infty}^{\infty} |g(y)|\, dy + |g(x)| \int_{|z| \geq \delta/h_n} |K(z)|\, dz.$$

Now, as $n \to \infty$, since $h_n \to 0$ the second and third terms go to zero, since $g \in L^1$ and $\lim_{y \to \infty} |yK(y)| = 0$. Then, as $\delta \to 0$, the first term goes to zero since $K \in L^1$ and x is a point of continuity of g. ∎

Thus, we immediately have

Theorem 4. The kernel estimator $\hat{f}_n$ in (117) subject to (118) and (119) is asymptotically unbiased if $h_n \to 0$ as $n \to \infty$, i.e.,

$$\lim_n E(\hat{f}_n(x)) = f(x).$$

Proof. Merely observe that

(122)
$$E[\hat{f}_n(x)] = \frac{1}{h_n} \int K\left(\frac{x - x_j}{h_n}\right) f(x_j) \, dx_j$$
$$= \int K(y) f(x - h_n y) \, dy$$
$$= f(x) + 0(h_n). \quad ∎$$

More importantly, we have

Theorem 5. The estimator $\hat{f}_n$ in (117) subject to (118) and (119) is consistent if we add the additional constraint $\lim_{n \to \infty} nh_n \to \infty$.

Proof. We have

$$\text{Var}[\hat{f}_n(x)] = \frac{1}{n} \text{Var}\left[\frac{1}{h_n} K\left(\frac{x - y}{h_n}\right)\right]$$

(since $\text{Var}(\bar{z}) = \frac{1}{n} \text{Var}(z)$).

Now

(123)
$$\frac{1}{n} \text{Var}\left[\frac{1}{h_n} K\left(\frac{x - y}{h_n}\right)\right] \leq \frac{1}{n} E\left[\left(\frac{1}{h_n} K\left(\frac{x - y}{h_n}\right)\right)^2\right]$$
$$= \frac{1}{h_n n}\left[\frac{1}{h_n} \int K^2\left(\frac{x - y}{h_n}\right) f(y) \, dy\right]$$
$$\to 0 \quad \text{if} \quad \lim_{n \to \infty} h_n n = \infty.$$

But

(124)
$$\text{MSE}[\hat{f}_n(x)] = E[(\hat{f}_n(x) - f(x))^2]$$
$$= \text{Var}[\hat{f}_n(x)] + \text{Bias}^2(\hat{f}_n(x)).$$

We have just proved that the variance goes to zero if $\lim\limits_{n\to\infty} h_n n = \infty$. That the bias goes to zero was proved in the preceding theorem. Thus, $\text{MSE}[\hat{f}_n(x)] \to 0$, i.e., $\hat{f}_n(x)$ is a consistent estimator for $f(x)$. ∎

We have already seen that for the Rosenblatt shifted histogram the optimal rate of decrease of the MSE is of the order of $n^{-4/5}$. We now address the question of the degree of improvement possible with the Parzen family of kernel estimators. Let us consider the Fourier transform of the kernel K, where K satisfies the conditions in (118) and (119). That is, let

$$(125) \qquad k(u) = \int_{-\infty}^{\infty} e^{iuy} K(y)\, dy,$$

assuming k is absolutely integrable.
Then, letting

$$(126) \qquad \varphi_n(u) = \int_{-\infty}^{\infty} e^{iux}\, dF_n(x) = \frac{1}{n} \sum_{k=1}^{n} e^{iux_k},$$

we have

$$(127) \qquad \hat{f}_n(x) = \frac{1}{nh_n} \sum_{k=1}^{n} K\{(x - x_k)/h_n\}$$

$$= \frac{1}{2\pi} \int e^{-iux} k(h_n u)\varphi_n(u)\, du$$

by the Levy Inversion Theorem.
Now,

$$(128) \qquad E[\varphi_n(x)] = \frac{1}{n} \sum_{k=1}^{n} E(e^{iux_k})$$

$$= \varphi(u),$$

where

$$\varphi(u) = E(e^{iux}).$$

Thus,

$$(129) \qquad E[\hat{f}_n(x)] = \frac{1}{2\pi} \int_{-\infty}^{\infty} e^{-iux} k(h_n u)\varphi(u)\, du.$$

So

$$(130) \qquad \text{Bias}[\hat{f}_n(x)] = \frac{1}{2\pi} \int_{-\infty}^{\infty} e^{-iux}[k(h_n u) - 1]\varphi(u)\, du.$$

Now, if there exists a positive r, such that

(131)
$$k_r = \lim_{u \to 0} \left[\frac{1 - k(u)}{|u|^r} \right]$$

is nonzero and finite, it is called the *characteristic exponent* of the transform k, and k_r is called the *characteristic coefficient*. But

(132) $\quad k_r = \lim_{u \to 0} \left[\frac{1 - k(u)}{|u|^r} \right] = \lim_{u \to 0} \left[\frac{1 - \int_{-\infty}^{\infty} e^{iux} K(x)\, dx}{|u|^r} \right]$

$$= \lim_{u \to 0} \frac{1}{|u|^r} \left[1 - \int_{-\infty}^{\infty} K(x)\, dx - iu \int_{-\infty}^{\infty} x K(x)\, dx - \cdots \right.$$

$$\left. - \frac{(iu)^{r-1}}{(r-1)!} \int_{-\infty}^{\infty} x^{r-1} K(x)\, dx - \frac{(iu)^r}{r!} \int_{-\infty}^{\infty} x^r K(x)\, dx + 0(u^{r+1}) \right].$$

Clearly, if k_r is to be finite and positive, we must have

(133)
$$\begin{cases} \int K(x)\, dx = 1 \\[2mm] \int_{-\infty}^{\infty} x^j K(x)\, dx = 0 \quad \text{for} \quad j = 1, 2, \ldots, r - 1 \\[2mm] \int x^r K(x)\, dx \neq 0 \quad \text{or} \quad \infty. \end{cases}$$

If K is itself to be a probability density, and therefore nonnegative, $r > 2$ is impossible if we continue to require the conditions of (119). Clearly, $r = 2$ is the most important case. Examples of kernels with characteristic exponents of 2 include the Gaussian, double exponential, and indeed any symmetric density K having $x^2 K(x) \in L^1$.

For the general case of a characteristic exponent r,

(134) $\quad \dfrac{\text{Bias}[\hat{f}_n(x)]}{h_n^r} = \dfrac{1}{2\pi} \int_{-\infty}^{\infty} e^{-iux} \dfrac{k(h_n u) - 1}{|h_n u|^r} |u|^r \varphi(u)\, du \to k_r f^{(r)}(x),$

where

(135)
$$f^{(r)}(x) = \frac{d^r f}{dx^r}\Big|_x = -\frac{1}{2\pi} \int_{-\infty}^{\infty} e^{-iux} |u|^r \varphi(u)\, du.$$

But from (123) we have

$$\text{Var}[\hat{f}_n(x)] \sim \frac{f(x)}{nh_n} \int K^2(y)\, dy.$$

Thus,

(136) $$\mathrm{MSE}[\hat{f}_n(x)] \sim \frac{f(x)}{nh_n} \int_{-\infty}^{\infty} K^2(y)\, dy + h_n^{2r} k_r^2 |f^{(r)}(x)|^2.$$

We may now, given f and K (and hence r), easily find that h_n which minimizes the asymptotic mean square error by solving

(137) $$2rh_n^{2r-1} k_r^2 |f^{(r)}(x)|^2 = \frac{f(x)}{nh_n^2} \int_{-\infty}^{\infty} K^2(y)\, dy,$$

giving

(138) $$h_n(x) = \left[f(x) \int_{-\infty}^{\infty} K^2(y)\, dy \middle/ \{ n2rk_r^2 |f^{(r)}(x)|^2 \} \right]^{1/(2r+1)}.$$

For this h_n, we have

(139) $$\mathrm{MSE}_{\mathrm{opt}} \sim (2r+1) \left\{ \frac{f(x)}{n2r} \int K^2(y)\, dy \right\}^{2r/(2r+1)} |k_r f^{(r)}(x)|^{2/(2r+1)}.$$

Thus we note that in the class of Parzen kernel estimators satisfying (118) the rate of decrease of the mean square error is of order $n^{-2r/(2r+1)}$. In practice, then, we should not expect a more rapid decrease of MSE for estimators of this class than $n^{-4/5}$—that obtainable by use of the shifted histogram.

A customary procedure is to obtain a global optimal h_n by minimizing the integrated mean square error

(140) $$\mathrm{IMSE}[\hat{f}_n] \sim \frac{1}{nh_n} \int_{-\infty}^{\infty} K^2(y)\, dy + h_n^{2r} k_r^2 \int_{-\infty}^{\infty} |f^{(r)}(x)|^2\, dx$$

to obtain

(141) $$h_n = n^{-1/(2r+1)} \left[\frac{\int K^2(y)\, dy}{2rk_r^2} \right]^{1/(2r+1)} \left[\int |f^{(r)}(y)|^2\, dy \right]^{-1/(2r+1)}$$

$$= n^{-1/(2r+1)} \left[\frac{\int K^2(y)\, dy}{2r \left(\int y^r K(y)\, dy/r! \right)^2} \right]^{1/(2r+1)} \left[\int |f^{(r)}(y)|^2\, dy \right]^{-1/(2r+1)}$$

$$= n^{-1/(2r+1)} \alpha(K) \beta(f).$$

Since the Parzen approach assumes the functional form of K (and hence r) to be given, the evaluation of $n^{-1/(2r+1)} \alpha(K)$ is unlikely to cause as much difficulty as the determination of $\beta(f)$.

In the following, we shall generally restrict our considerations to kernels where $r = 2$. Below we show a table of useful kernels with their standard supports and their corresponding α values

Table 2.1

K_i		$\alpha(K_i)$
$K_1(y) = \dfrac{1}{2}$	$\|y\| \leq 1$	1.3510
$K_2(y) = 1 - \|y\|$	$\|y\| \leq 1$	1.8882
$K_3(y) = \dfrac{15}{16}(1 - y^2)^2$	$\|y\| \leq 1$	2.0362
$K_4(y) = \dfrac{1}{\sqrt{2\pi}} e^{-y^2/2}$	$\|y\| < \infty$	0.7764

Although the Gaussian kernel K_4 is fairly popular, the quartic kernel estimator using K_3 is nearly indistinguishable in its smoothness properties and has distinct computational advantages due to its finite support.

In Figures 2.1 through 2.6, we display an empirical interactive technique for picking h_n using K_3. A random sample of size 300 has been generated from the bimodal mixture

$$(142) \quad f(x) = .75 \frac{1}{\sqrt{2\pi}} \exp\left[-\frac{(x + 1.5)^2}{2}\right] + .25 \frac{3}{\sqrt{2\pi}} \exp\left[-\frac{(x - 1.5)^2}{2/9}\right].$$

We note that for $h = 5$, the estimated density is overly smoothed. However, the user who does not know the functional form of f would not yet realize that the estimator is unsatisfactory. Now, as h is decreased to 2, the bimodality of f is hinted. Continuing the decrease of h, the shape of f becomes clearer. When we take h to .2, the noise level of the estimator becomes unacceptable. Such additional structure in f beyond that which we have already perceived for slightly larger h_n values is obscured by the fact that nh_n is too small. Accordingly, the user would return to the estimator obtained for $h_n = .4$.

We note here the fact that kernel estimators are not, in general, robust against poor choices of h_n. A guess of h_n a factor of 2 from the optimal may frequently double the integrated mean square error. Accordingly, we highly recommend the aforementioned interactive approach, where the user starts off with h_n values which are too large and then sequentially decreases h_n until overly noisy probability density estimates are obtained. The point where

Figure 2.1. $n = 300$ bimodal quartic kernel $h_n = 5.0$.

Figure 2.2. $n = 300$ bimodal quartic kernel $h_n = 2.0$.

Figure 2.3. $n = 300$ bimodal quartic kernel $h_n = 0.8$.

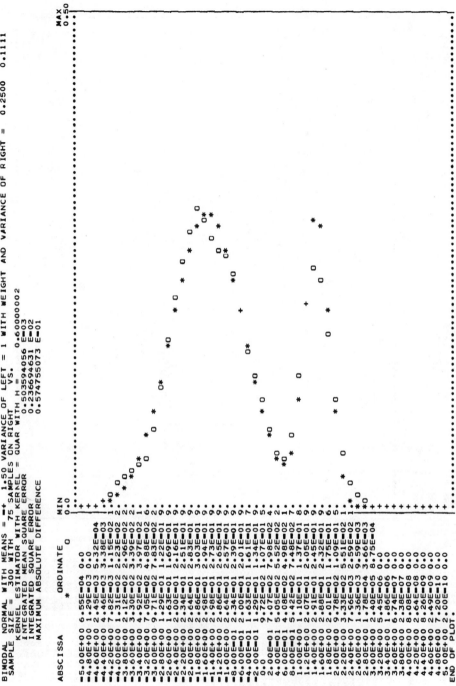

Figure 2.4. $n = 300$ bimodal quartic kernel $h_n = 0.6$.

Figure 2.5. $n = 300$ bimodal quartic kernel $h_n = 0.4$.

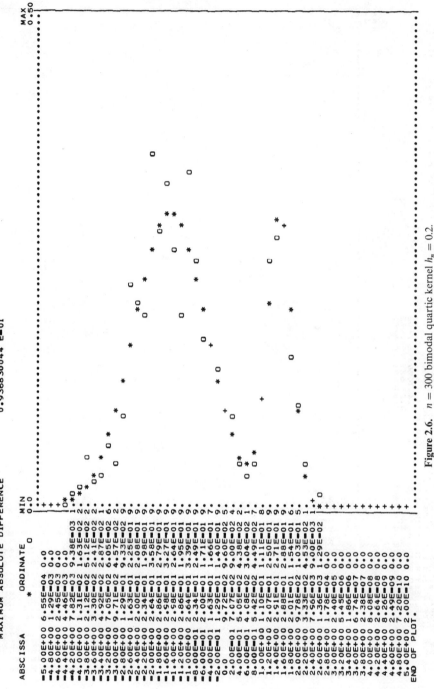

Figure 2.6. $n = 300$ bimodal quartic kernel $h_n = 0.2$.

further attempts at resolution by decreasing h_n drives one into a highly noisy estimate is generally fairly sharp and readily observable. However, for the user who is not really into time sharing, we give below an empirical algorithm which is frequently successful in selecting a satisfactory value of h_n, using a batch, i.e., one shot, approach.

Did we but know $\int_{-\infty}^{\infty} (f''(y))^2 \, dy$, there would be no trouble in picking a good h_n for a particular kernel with characteristic exponent $r = 2$. It is tempting, since $\int_{-\infty}^{\infty} (f''(y))^2 \, dy$ is sensitive, though not terribly so, to changes in f, to try functional iteration to obtain a satisfactory value for h_n. Thus, if we have a guess for h_n—say h_n^i—we may use (127) to obtain an estimate—say $\hat{f}_n^i$—for f. Then from (141), we have as our iterated guess

$$(143) \qquad h_n^{i+1} = n^{-1/5}\alpha(K)\beta(\hat{f}_n^i).$$

More efficiently, since β is an explicit function of h_n, we may use Newton's method to find a zero of (143). Typically, we find that if Newton's method is employed, around five iterations are required to have $|h_n^{i+1} - h_n^i| < 10^{-5}$.

In Tables 2.2 and 2.3, we demonstrate the results of Monte Carlo simulations using (143). The distributions examined were the standard normal, the 50–50 mixture of two normal densities with a separation of means of three times their common standard deviation, a central t distribution with 5 degrees of freedom, and an $\mathscr{F}$ distribution with (10, 10) degrees of freedom. The average values of h_n^∞, together with the theoretically (asymptotically) optimal values of h_n, were computed from (143). In Table 2.3 we show integrated mean square error results obtained using the recursive technique in comparison with the kernel estimator using the theoretically optimal value of h_n. The average efficiency shown is simply the average of the ratio of the IMSE obtained using the theoretically optimal value of h_n divided

Table 2.2   Monte Carlo Results for Finding the Quasi-Optimal h_n
(Each Row Represents 25 Samples)

Density	Sample size	Degenerate*	Mean	Std. dev.	Range	Theoretical
$N(0, 1)$	25	1	.54	.17	.20–.80	.56
$.5N(-1.5, 1) + .5N(1.5, 1)$	25	2	.77	.41	.09–1.35	.66
t_5	25	0	.59	.19	.25–.96	.41
$F_{10,10}$	25	2	.25	.09	.02–.42	.20
$N(0, 1)$	100	0	.35	.10	.09–.51	.42
$.5N(-1.5, 1) + .5N(1.5, 1)$	100	0	.43	.17	.12–.76	.50
t_5	100	0	.37	.09	.13–.54	.31
$F_{10,10}$	100	1	.15	.04	.05–.20	.15

* In less than 1% of the cases examined $h_n^\infty = 0$. These cases were excluded from the study.

Table 2.3   Integrated Mean Square Error, Using the Quasi-Optimal h_n
vs. the Theoretically Optimal h

Density	Number of samples	Sample size	Quasi-optimal h_n Mean	Std. dev.	Theoretical h_n Mean	Std. dev.	Average efficiency (%)
$N(0, 1)$	24	25	.0242	.0187	.0163	.119	67
$.5N(-1.5, 1) + .5N(1.5, 1)$	22*	25	.0233	.0306	.0095	.0070	41
t_5	25	25	.0230	.0109	.0210	.0091	91
$F_{10,10}$	22*	25	.0401	.0194	.0390	.0187	97
$N(0, 1)$	24*	100	.0071	.0042	.0050	.0028	70
$.5N(-1.5, 1) + .5N(1.5, 1)$	23*	100	.0053	.0024	.0036	.0021	68
t_5	25	100	.0085	.0099	.0069	.0039	81
$F_{10,10}$	24	100	.0217	.0221	.0168	.0094	77

* The largest one or two IMSE values were omitted because the corresponding quasi-optimal h_n was nearly zero.

by that obtained using the quasi-optimal value obtained by functional iteration.

Early on (1958) in the development of kernel density estimators, Whittle [60] proposed a Bayesian approach. Let us consider an estimator of the form

$$(144) \qquad \hat{f}_n(x) = \frac{1}{n} \sum_{j=1}^{n} K_x(x_j).$$

To develop a somewhat more concise set of expressions, Whittle considers the sample size n to be a Poisson variate with expectation M. Letting

$$(145) \qquad \varphi(x) = Mf(x),$$

we shall estimate $\varphi(x)$ with statistics, of the form

$$(146) \qquad \hat{\varphi}(x) = \sum_{j=1}^{n} K_x(x_j).$$

Whittle assumes prior values on $\varphi(x)$ and $\varphi(x)\varphi(y)$ via

$$(147) \qquad E_p[\varphi(x)] = \mu(x)$$

$$(148) \qquad E_p[\varphi(x)\varphi(y)] = \mu(x, y).$$

The assumption of the existence of an appropriate measure on the space of densities is a nontrivial one [32].

Then a straightforward calculation shows that the Bayes risk is minimized when

$$(149) \qquad \mu(y)K_x(y) + \int \mu(y, z)K_x(z)\, dz = \mu(y, x).$$

We then carry out the following normalization

$$(150) \qquad \xi_x(y) = K_x(y) \sqrt{\frac{\mu(y)}{\mu(x)}}$$

$$(151) \qquad \gamma(x, y) = \frac{\mu(x, y)}{\sqrt{\mu(x)\mu(y)}}.$$

Then (149) becomes

$$(152) \qquad \xi_x(y) + \int \gamma(y, z)\xi_x(z)\, dz = \gamma(y, x).$$

Whittle suggests special emphasis on the case where $\gamma(y, x) = \gamma(y - x)$. This property, obviously motivated by second-order stationarity in time series analysis, might be appropriate, Whittle argues, in the case where $\mu(x)$ is rectangular. Then (152) becomes

$$(153) \qquad \xi_x(y) + \int \gamma(y - z)\xi_x(z)\, dz = \gamma(y - x).$$

Now, if $\iint |\gamma(y - x)|^2 \, dx \, dy < \infty$, then (153) has a unique solution $\xi_x(y)$ in $L^2(-\infty, \infty)$, since $(I + \gamma*)$ is a Hilbert–Schmidt compact operator. Whittle solves for $\xi_x(y)$ by Fourier transforming (153) to obtain

$$(154) \qquad \xi_x(y) = \frac{1}{2\pi} \int_{-\infty}^{\infty} e^{i\omega(y - x)} \frac{\Gamma(\omega)}{1 + \Gamma(\omega)}\, d\omega$$

where Γ is the Fourier transform of γ. He proposes the following form for γ

$$(155) \qquad \gamma(x) = v(a + be^{-c|x|}).$$

(This causes some difficulty, since $\gamma \notin L^1(-\infty, \infty)$ if $a \neq 0$, and hence the Fourier transform does not exist.)
Then, solving for $\xi_x(y)$ we have

$$(156) \qquad \xi_x(y) = \frac{vbc}{\theta} e^{-\theta|y - x|},$$

where

$$(157) \qquad \theta = (2vbc + c^2)^{1/2}.$$

But then $\xi_x(y)$ does not depend on a. Let us consider two possibilities for γ, namely,

$$\gamma_1(x) = v(a_1 + be^{-c|x|})$$

and

$$\gamma_2(x) = v(a_2 + be^{-c|x|}).$$

Both γ_1 and γ_2 yield the same $\xi_x(y)$. So from (153) we have

(158) $$(va_1 - va_2) \int \xi_x(z)\, dz = va_1 - va_2.$$

Thus, any solution to (153) must have

(159) $$\int \xi_x(z)\, dz = 1.$$

But this requires, from (156) and (157) that

$$\frac{2vbc}{\theta^2} = 1$$

or that

$$\frac{2vbc}{2vbc + c^2} = 1.$$

Thus, it must be true that $c^2 = 0$. Hence, it must be true that $\theta = 0$. Thus, we have shown

Theorem 6. (156) is not a solution to (153) for arbitrary a. In fact, it is immediately clear that in order for (156) to solve (153) we must have $a = 0$ (which, of course, makes γ an $L^1(-\infty, \infty)$ function).

We can also establish the "translation invariance" of $\xi_x(y)$, namely,

Theorem 7. $\xi_x(y) = \xi_0(y - x)$.

Proof.

$$\xi_x(y) = \frac{vbc}{\theta}\, e^{-\theta|y-x|} = \xi_0(y - x).$$

(This theorem is true for any second-order stationary γ.) ∎

Now, to return to the original problem of obtaining a (Bayesian) optimal form for $K_x(y)$, we use (150) and (156) to obtain

(160) $$K_x(y) = \sqrt{\frac{\mu(x)}{\mu(y)}}\, \frac{vbc}{\theta}\, e^{-\theta|y-x|}.$$

But if $\hat{f}_n(x)$ is to be a probability density, we must have

(161) $$1 = \int \hat{f}_n(x)\, dx = \frac{1}{n} \sum_{j=1}^{n} \int K_x(x_j)\, dx.$$

For the second-order stationarity assumption $\gamma(y, x) = \gamma(y - x)$ to be reasonable, Whittle suggests that one should assume $\mu(x) = \mu(y)$ for all (x, y). But this would imply

(162) $$K_x(y) = \frac{vbc}{\theta}\, e^{-\theta|y-x|}.$$

and that

(163)
$$\frac{1}{n} \sum_{j=1}^{n} \frac{vbc}{\theta} \int e^{-\theta|x_j - x|} \, dx = 1.$$

If this is to hold for any possible set of sample points $\{x_j\}_{j=1}^{n}$, we must have

(164)
$$\frac{2vbc}{\theta^2} = 1.$$

We have seen already that this is impossible. Thus, if (161) is to hold, the kernel in (162) gives an empty family of estimators.

Now if γ is any second-order stationary kernel, we are led to estimates of the form

$$\hat{f}_n(x) = \frac{1}{n} \sum_{j=1}^{n} \frac{\mu(x)}{\mu(x_j)} \xi_0(x_j - x).$$

If this is to be a density for any possible sample $\{x_j\}_{j=1}^{n}$, we must have ξ_0 to be a probability density and $\mu(x) = \mu(y)$ for all x and y. Thus, after much labor, we are led, simply, to a Parzen estimator. A more recent data-oriented approach is given by de Figueiredo [17].

If one finds it unreasonable or impractical to make rather strong prior assumptions about the unknown density f, it is tempting to try a minimax approach for optimal kernel design. Such an attack has been carried out by Wahba [53] motivated by her earlier paper [52] and a result of Farrell [15].

Let us assume that $f \in W_p^{(m)}(M)$, i.e., $[\int_{-\infty}^{\infty} |f^{(m)}(x)|^p \, dx]^{1/p} \le M$, where $p \ge 1$.

Since f is a density, there exists a $\Lambda < \infty$ such that $\sup_{-\infty < x < \infty} f(x) \le \Lambda$.

Next, we shall restrict out consideration to kernels which satisfy the following conditions:

(165)
$$\sup_{-\infty < y < \infty} |K(y)| < \infty$$

(166)
$$\int_{-\infty}^{\infty} |K(y)| \, dy < \infty$$

(167)
$$\lim_{|y| \to \infty} |yK(y)| = 0$$

(168)
$$\int_{-\infty}^{\infty} K(y) \, dy = 1$$

(169)
$$\int_{-\infty}^{\infty} y^j K(y) \, dy = 0, \qquad j = 1, 2, \ldots, m - 1$$

(170)
$$\int_{-\infty}^{\infty} |y|^{m - 1/p} |K(y)| \, dy < \infty.$$

Now, for the estimators of the form

$$\hat{f}_n(x) = \frac{1}{nh_n} \sum_{k=1}^{n} K\left(\frac{x - x_k}{h_n}\right),$$

we have

(171) $\quad\quad \text{Var}[\hat{f}_n(x)] = \frac{f(x)}{nh_n} \int_{-\infty}^{\infty} K^2(y)\, dy \left[1 + 0\left(\frac{1}{nh_n}\right)\right]$

$$\leq B \frac{1}{nh_n}\left[1 + 0\left(\frac{1}{nh_n}\right)\right],$$

where

$$B = \Lambda \int_{-\infty}^{\infty} K^2(y)\, dy.$$

The bias term is given by

(172) $\quad \text{Bias}[\hat{f}_n(x)] = E[\hat{f}_n(x)] - f(x) = \int_{-\infty}^{\infty} K(-\xi)f(x + \xi h)\, d\xi - f(x).$

Taking a Taylor's expansion for $f(x + \xi h)$ we have

(173) $\quad f(x + \xi h) = f(x) + \sum_{i=1}^{m-1} \frac{(\xi h)^i}{i!} f^{(i)}(x) + \int_{x}^{x + \xi h} \frac{(x + \xi h - u)^{m-1}}{(m-1)!} f^{(m)}(u)\, du.$

Using (168), (169), (170), and (173) in (172), we have

(174) $\quad E[\hat{f}_n(x)] - f(x) = \int_{-\infty}^{\infty} K(-\xi) \int_{x}^{x + \xi h} \frac{(x + \xi h - u)^{m-1}}{(m-1)!} f^{(m)}(u)\, du\, d\xi.$

Applying Hölder's inequality to the inner integral in (174), we have

(175) $\quad\quad\quad\quad \text{Bias}^2[\hat{f}_n(x)] \leq M^2 A h_n^{2m - 2/p},$

where

(176) $\quad A = \dfrac{1}{[(m-1)!]^2[(m-1)q + 1]^{2/q}} \left[\int_{-\infty}^{\infty} |K(\xi)|\, |\xi|^{m - 1/p}\, d\xi\right]^2,$

with

$$\frac{1}{q} + \frac{1}{p} = 1.$$

Thus, ignoring the factor $\left(1 + 0\left(\dfrac{1}{nh_n}\right)\right)$ we have

(177) $\quad\quad\quad\quad \text{MSE}(\hat{f}_n(x)) \leq M^2 A h_n^{2m - 2/p} + B \frac{1}{nh_n}.$

The right-hand side of (177) is minimized for

$$(178) \qquad h_n = \left[\frac{1}{2m - 2/p} \frac{B}{M^2 A} \right]^{1/(2m+1-2/p)} n^{-1/(2m+1-2/p)}.$$

Substituting this value back into (177), we have

$$(179) \qquad \text{MSE}(\hat{f}_n(x)) \leq Dn^{-(2m-2/p)/(2m+1-2/p)}(1 + o(1)),$$

where

$$D = \frac{(2m + 1 - 2/p)}{(2m - 2/p)^{(2m-2/p)}} (M^2 AB^{2m-2/p})^{1/(2m+1-2/p)}$$

and

$$AB^{2m-2/p} = \frac{1}{[(m - 1)!\{((m - 1)/(1 - 1/p)) + 1\}^{1 - 1/p}]^2}$$

$$\times \left[\int_{-\infty}^{\infty} |K(\xi)| \, |\xi|^{m-1/p} \, d\xi \right]^2 \left[\Lambda \int_{-\infty}^{\infty} K^2(y) \, dy \right]^{2m-2/p}.$$

Thus, we know that for any density in $W_p^{(m)}(M)$ the mean square error is no greater than the right-hand side of (179). We do not know how tight this upper bound on the mean square error is. However, Wahba observes that, on the basis of the bound, we should pick K so as to minimize

$$(180) \qquad \left[\int_{-\infty}^{\infty} |K(\xi)| \, |\xi|^{m-1/p} \, d\xi \right] \left[\int_{-\infty}^{\infty} K^2(\xi) \, d\xi \right]^{m-1/p}.$$

This task has been undertaken by Kazakos [26], who notes that the upper bound is achieved for $W_p^2(M)$ by

$$(181) \qquad K(y) = \frac{(1 + a^{-1})}{2}[1 - |y|^a] \quad \text{for} \quad |y| < 1$$

$$= 0 \quad \text{otherwise},$$

where $a = 2 - 1/p$.

(We have here omitted a superfluous scale parameter used by Kazakos. The special case where $p = \infty$ gives the quadratic window of Epanechnikov [12] discussed also by Rosenblatt [39].)

We can now evaluate A, B, h_n and the Wahba upper bound on the mean square error:

$$(182) \qquad A = \frac{(1 + a^{-1})^2}{[(2 - 1)!]^2[(2 - 1)q + 1]^{2/q}} \left[\int_0^1 (1 - y^a)y^{2-1/p} \, dy \right]^2$$

$$= \frac{1}{(q + 1)^{2/q}(2a + 1)^2}$$

(183) $\qquad B = \Lambda \dfrac{(1 + a^{-1})^2}{2} \displaystyle\int_0^1 (1 - 2y^a + y^{2a})\, dy$

$$= \Lambda \frac{a + 1}{2a + 1}.$$

h_n and the Wahba upper bound on the MSE may be obtained by substituting for A and B in (178) and (179) respectively.

Returning to (178), let us consider

(184) $\qquad h_n = \left[\dfrac{(q + 1)^{2/q}}{4 - 2/p} \dfrac{\displaystyle\int K^2(y)\, dy}{\left[\displaystyle\int |K(y)|\, |y|^{2 - 1/p}\, dy \right]^2} \dfrac{\Lambda}{M^2} \right]^{1/(5 - 2/p)} n^{-1/(5 - 2/p)}.$

In most practical cases, if f has infinite support it will be reasonable to assume that $f^{(2)}$ is bounded. To obtain the fastest rate of decrease in n, we shall take $p = \infty$, thus giving

(185) $\qquad h_n^{(1)} = \left[\dfrac{\displaystyle\int K^2(y)\, dy}{\left[\displaystyle\int |K(y)| y^2\, dy \right]^2} \right]^{1/5} \left[\dfrac{\sup |f^{(2)}|^2}{\sup\limits_x f(x)} \right]^{-1/5} n^{-1/5}$

$$\left(= \left(\frac{15\Lambda}{M^2} \right)^{1/5} n^{-1/5} \text{ (using the kernel in (181))} \right).$$

Comparing (185) with the h_n obtained from the integrated mean square error criterion formula in (141), i.e.,

(186) $\qquad h_n^{(2)} = \left[\dfrac{\displaystyle\int K^2(y)\, dy}{\left[\displaystyle\int |K(y)| y^2\, dy \right]^2} \right]^{1/5} \left[\displaystyle\int |f^{(2)}(y)|^2\, dy \right]^{-1/5} n^{-1/5}.$

We note that the term in the first brackets of (185) and (186) are identical. The only difference is in the second bracket term. How reasonable is it to suppose that we will have reliable knowledge of $\sup |f^{(2)}|$ and $\sup |f|$? Probably it is not frequently the case that we shall have very good guesses as to these quantities. Of course, an iterative algorithm could be constructed to estimate these quantities from the data. However, data-based estimates for these two quantities violate the philosophy of minimax kernel construction. These suprema in (185) are taken with respect to all densities in $W_\infty^{(2)}(M)$. Even if we were content to replace these suprema using a kernel estimator based on the data, we would, of course, find them unreliable relative to the earlier mentioned estimate (143) for $\int (f^{(2)})^2 \, dx$. Moreover, the kernels given in (181) suffer from rough edges. Scott's quartic kernel K_3 [43], a form of Tukey's biweight function [see §2.3], given in Table 2.1, does not have this difficulty.

It is intuitively clear that if we actually know the functional form of the density f, a kernel estimator will be inferior to that obtained by the classical method of using standard techniques for estimating the parameters which characterize f, and substituting these estimates for the parameters into f. For example, let us consider the problem for estimating the normal density with unit variance but unknown mean μ on the basis of a random sample of size n. If $\bar{x}$ is the sample mean, then the classical estimate for f is given by:

$$(187) \qquad \hat{f}_p = \frac{1}{\sqrt{2\pi}} \exp\left[-\frac{1}{2}(x - \bar{x})^2 \right].$$

Then the integrated mean square error is given by

$$(188) \qquad \text{IMSE} = \frac{1}{\Pi}\left[1 - \sqrt{\frac{n}{n + \frac{1}{2}}} \right]$$

$$= \frac{1}{\Pi}\left[\frac{1}{4(n + \frac{1}{2})} + 0\left(\frac{1}{n^2}\right) \right].$$

Now, from (140) and (141) we have that the integrated mean square error of a kernel estimator based on the (density) kernel K of characteristic exponent 2 is bounded below by

$$(189) \quad \text{IMSE} \geq \frac{5}{4}\left[\int K^2(y) \, dy \right]^{4/5} \left[\int y^2 K(y) \, dy \right]^{2/5} \left[\int (f''(y))^2 \, dy \right]^{1/5} n^{-4/5}.$$

We can attain this bound only if we actually know $\int |f''(y)|^2 \, dy$. We have seen that using our suggested procedure in (143) for estimating $\int |f''(y)|^2 \, dy$ actually increases IMSE values over this bound significantly. Nonetheless, let us examine the efficiency of the kernel estimator in the utopian case where we actually know $\int |f''(y)|^2 \, dy$. We shall first use the Gaussian kernel K_4. In this case, we have

$$(190) \qquad \text{IMSE} \geq .3329 n^{-4/5}.$$

Thus, the Gaussian kernel estimator has an efficiency relative to the classical estimator, which gradually deteriorates to zero as sample size goes to infinity. However, for a sample of size 100, the efficiency bound is 17%; and it has only fallen to 13% when the sample size has increased to 400. Moreover, we must remember that a nonparametric density estimation technique should not be used if we really know the functional form of the unknown density. If the true form of the density is anything (having $f'' \in L^2$) other than $N(\mu, 1)$, then the efficiency of the Gaussian kernel estimator, relative to the classical estimator, assuming the true density is $N(\mu, 1)$, becomes infinite as the sample size goes to infinity.

Nevertheless, there has been great interest in picking kernels which give improved efficiencies relative to the classical techniques which assume the functional form of the density is known. In particular, there has been extensive work on attempts to bring the rate of decrease of the IMSE as close to n^{-1} as possible. The first kernel estimator, the shifted histogram of Rosenblatt [38], showed a rate of decrease of $n^{-4/5}$.

Looking at (189), one notices that if one concedes the practical expedient of using kernels with IMSE rate of decrease $n^{-4/5}$, it is still possible to select K so as to minimize $[\int K^2(y) \, dy]^{4/5} [\int y^2 K(y) \, dy]^{2/5}$. A straightforward variational argument [12] shows that this may be done (in the class of symmetric density kernels) by using

$$(191) \qquad K_5(y) = 3/4(1 - y^2) \quad \text{if} \quad |y| \leq 1$$
$$= 0 \quad \text{otherwise.}$$

This is, as stated earlier, the kernel obtained in (181) by considering the Wahba "lower bound" on the mean square error for $m = 2$, $p = \infty$. Then the lower bound on the integrated mean square is given by

$$(192) \qquad \text{IMSE} \geq .3198 n^{-4/5}.$$

But comparing (192) with (190) we note that the Epanechnikov kernel estimator gives only a 4% improvement in the lower bound over the Gaussian kernel estimator. As a matter of fact, all the practical kernel estimators with characteristic exponent 2 have lower bound IMSE efficiencies very close to that of the "optimal" Epanechnikov kernel. We note that the non-differentiability of the Epanechnikov kernel at the endpoints may cause practical difficulties when one faces the real world situation where $\int |f''(x)|^2 \, dx$ must be estimated from the data. The Gaussian kernel does not suffer from this difficulty, but it has the computational disadvantage of infinite support. A suitable compromise might be K_3 or the B-spline kernel [2].

Among the first attempts to develop kernels which have IMSE decrease close to the order of n^{-1} was the 1963 work of Watson and Leadbetter [55]. They sought to discover that estimator

$$(193) \qquad \hat{f}(x) = \frac{1}{n} \sum_{j=1}^{n} K_n(x - x_i),$$

which minimized

(194)
$$E\left[\int (\hat{f}_n(x) - f(x))^2\, dx\right].$$

As a first step, they assumed that one knows the density $f(x)$ *precisely* insofar as computation of K is concerned. A straightforward application of Parseval's theorem yields as the formula for the Fourier transform of K_n:

(195)
$$\varphi_{K_n}(t) = \frac{|\varphi_f(t)|^2}{\dfrac{1}{n} + \left[\dfrac{n-1}{n}\right]|\varphi_f(t)|^2},$$

where φ_f is the characteristic function of the density f. We note that the constraint that K_n be a density has not been applied, and the K_n corresponding to $\varphi_{K_n}(t)$ need not be a probability density. Consequently, estimated densities using the Watson–Leadbetter procedure need not be nonnegative nor integrate to unity. However, the procedure provides lower bounds for the integrated mean square error. Certainly, one should not expect to do better in kernel construction (purely by the IMSE criterion) than the case where one assumes the true density to be known and relaxes the constraint that a kernel be a density.

Watson and Leadbetter attempt to bring their procedure into the domain of practical application by considering situations where one may not know φ_f precisely. Let it be known that

(196) $\displaystyle\lim_{|t|\to\infty} |t|^p |\varphi_f(t)| = C^{1/2} > 0$ with $p > \frac{1}{2}$, i.e., that the characteristic

function of f decreases algebraically of degree $p > \frac{1}{2}$. Then,

(197)
$$\hat{\varphi}_{K_n}(t) = h(A_n t),$$

where

$$h(t) = (1 + |t|^{2p})^{-1}$$
$$A_n = [nC]^{-1/2p}$$

gives us strictly positive asymptotic efficiency relative to the estimator given in (195). The order of decrease of the IMSE of the kernel estimator using $\hat{\varphi}_{K_n}$ is $n^{-1+2/p}$. Unfortunately, the resulting estimator is not a probability density.

Next, suppose $\varphi_f(t)$ has exponential decrease of degree 1 and coefficient ρ, i.e.,

(198)
$$|\varphi_f(t)| \le A \exp[-\rho|t|^1]$$

for some constant A and all t, and

(199)
$$\lim_{v \to \infty} \int_0^1 [1 + e^{2\rho v}|\varphi_f(vt)|^2]^{-1} \, dt = 0.$$

Then, if

$$\hat{\varphi}_{K_n}(t) = h(A_n e^{\alpha|t|}),$$

where

$$h \in L^2$$

$$\sup_t |h(t)| < B < \infty$$

$$|1 - h(t)| \le B_1|t| \quad \text{for} \quad |t| \le 1$$
$$A_n = Dn^{-b}$$
$$\alpha = 2\rho b$$
$$b > \tfrac{1}{2},$$

then the resulting kernel estimator has a positive asymptotic efficiency relative to the estimator given in (195). The rate of decrease of the integrated mean square error is $(\log n)n^{-1}$.

A related approach of Davis [11] based on "saturation" theorems of Shapiro [45] seeks to improve the rate of convergence of the mean square error by eliminating the requirement that $K \in L^1$ from the conditions in (118). We recall from (134) that for $K \in L^1$

(200)
$$\frac{\text{Bias}[\hat{f}_n(x)]}{h_n^r} = \frac{1}{2\Pi} \int_{-\infty}^{\infty} e^{-iux} \frac{k(h_n u)^{-1}}{|h_n u|^r} |u|^r \varphi(u) \, du \to k_r f^{(r)}(x),$$

where k_r is the characteristic coefficient of K and r is its characteristic exponent.

Now, for all $j = 1, 2, \ldots, r - 1$

(201)
$$\int_{-\infty}^{\infty} x^j K(x) \, dx = 0.$$

If we require that the conditions of (118) be operative, then we can do no better than $r = 2$. But then we can do no better than

(202)
$$\text{Bias}^2[\hat{f}_n(x)] = 0(h_n^4).$$

But

(203)
$$\text{Var}[\hat{f}_n(x)] = 0\left(\frac{1}{nh_n}\right).$$

Thus, we are forced to sacrifice the rate of decrease in the variance in order to accommodate the Bias^2 term. By eliminating the condition that $K \in L^1$, we are enabled to consider kernels which do not have their MSE's decreasing so slowly as $n^{-4/5}$.

The suggested kernel of Davis is simply the sinc function

$$(204) \qquad\qquad K(x) = \frac{\sin x}{\Pi x}.$$

The sinc function is, of course, the Fourier transform of a step function. Generally, sinc functions occur in practice when one takes the Fourier transform of a signal limited in time extent. It is a great villain in time series analysis due to its heavy negative and positive minor lobes. However, in this context, Davis considers it to be a convenient non-L^1 (though L^2) function which integrates to unity. As we shall shortly see, the sinc function estimator can give substantial negative values. Of course, this is not surprising, since nonnegativity has neither explicitly nor implicitly been imposed as a condition.

In the case where the characteristic function of the unknown density shows algebraic decrease of order p, the sinc function estimator has Bias^2 decrease of order h_n^{2p-2}. The variance decrease is of order $1/(h_n n)$. Thus a choice of $h_n = cn^{-1(2p-1)}$ gives a mean square error decrease of order $n^{-1+1/(2p-1)}$.

In the case where the characteristic function of the density decreases exponentially with degree r and coefficient ρ, we have that Bias^2 of the sinc function estimator decreases at least as fast as $\exp[-2p/h_n^r]/h_n^2$. The variance decrease is still of order $1/nh_n$. A choice of $h_n = (\log n/2\rho)^{-1/r}$ gives mean square error decrease of order $(\log n)/n$, i.e., very close to the n^{-1} rate. In Figures 2.7 and 2.8 we show the results of applying the Davis algorithm to the estimation of $N(0, 1)$ for samples of size 20 and 100 respectively. The substantial negative side lobes are disturbing. This difficulty is even more strikingly seen in Figure 2.9, where we apply the sinc function procedure to estimation of an F density with $(10, 10)$ degrees of freedom on the basis of a sample of size 400.

It is very likely that the negativity of non-L^1 estimates would be lessened by using kernels like those constructed by Tukey for time series analysis, e.g.,

$$(205) \qquad K(x) = \frac{1}{\Pi}\left[.5\frac{\sin x}{x} + .25\frac{\sin(x - a_n)}{(x - a_n)} + .25\frac{\sin(x + a_n)}{(x + a_n)}\right].$$

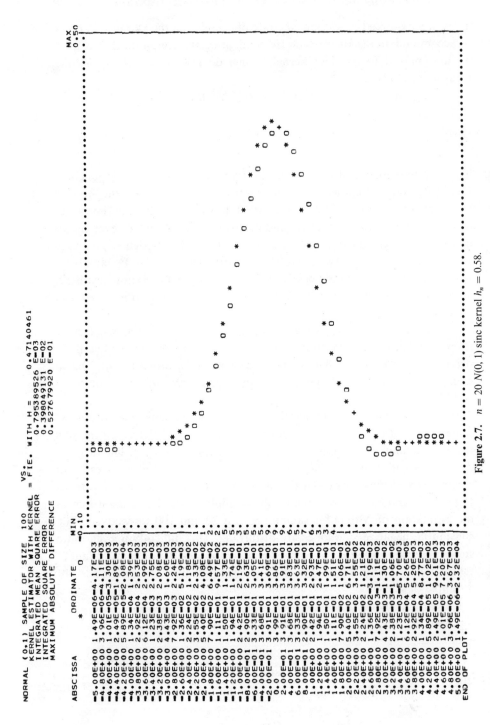

Figure 2.7. $n = 20$ $N(0, 1)$ sinc kernel $h_n = 0.58$.

NORMAL (0,1) SAMPLE OF SIZE 20   VS.
KERNEL ESTIMATOR WITH KERNEL =FIE. WITH H= 0.58447391
INTEGRATED MEAN SQUARE ERROR   0.153429294 E—02
INTEGRATED SQUARE ERROR   0.106731430 E—01
MAXIMUM ABSOLUTE DIFFERENCE   0.778741837 E—01

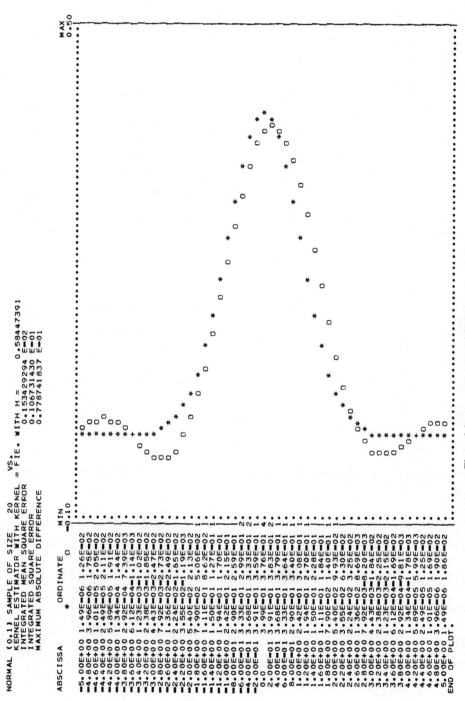

Figure 2.8. $n = 100$ $N(0, 1)$ sinc kernel $h_n = 0.47$.

81

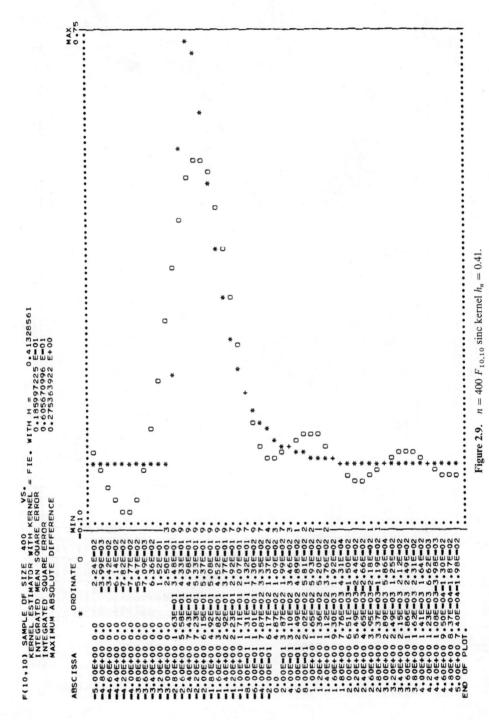

Figure 2.9. $n = 400$ $F_{10,10}$ sinc kernel $h_n = 0.41$.

In a sequence of papers [56, 57, 58, 59], Wegman has employed a maximum likelihood approach to obtain a modified histogram estimator for an unknown density f with domain $(a, b]$. His estimators are of the form

$$(206) \qquad f(x|c) = \sum_{j=0}^{k-1} c_j I_{(a_j, a_{j+1}]}^{(x)},$$

where

$$I_{(c,d]}^{(x)} = 1 \quad \text{if} \quad x \in (c, d]$$
$$= 0 \quad \text{otherwise}$$

$$c_j > 0 \qquad j = 0, 1, \ldots, k$$

$$\sum_{j=0}^{k-1} c_j(a_{j+1} - a_j) = 1$$

$$a_0 = a$$
$$a_k = b.$$

The criterion function to be maximized based on a sample of size n $\{x_1, x_2, \ldots, x_n\}$ is

$$(207) \qquad L(c) = \prod_{i=1}^{n} \hat{f}(x_i|c)$$

subject to the constraint that at least $m(n)$ observations must fall into each of the k intervals where $k \le [n/m(n)]$ and $m(n)$ goes to infinity faster than $0(\sqrt{\log(\log n)})$.

In the case where $n = km(n)$, the solution is particularly simple, being analogous to that given in Theorem 2. Thus we have

$$(208) \qquad \hat{f}(x) = \frac{1}{k} \sum_{j=0}^{k-1} \frac{1}{a_{j+1} - a_j} I_{(a_j, a_{j+1}]}^{(x)}$$

$$a_0 = a$$
$$a_1 = x_{(1m)}$$
$$a_2 = x_{(2m)}$$
$$\vdots$$
$$a_{k-1} = x_{(\{k-1\}m)}$$
$$a_k = b.$$

Effectively, then, Wegman suggests that the interval widths of the histogram should vary across the data base in a manner inversely proportional to the density of data points in the interval.

This approach has implications for the selection of h_{in} in the case of a kernel density estimator:

(209) $$\hat{f}_{in}(x) = \frac{1}{n} \sum_{i=1}^{n} K\{(x - x_i)/h_{in}\}/h_{in}.$$

We might, for example, use

$$h_{in}^{k} = \text{distance of } x_i \text{ to the } k^{th} \text{ nearest observation.}$$

Clearly, k should vary with n. Perhaps

(210) $$k(n) = n^{1-\alpha}p,$$

where $0 < \alpha < 1$ and $0 < p < 1$
would be appropriate. For Parzen kernels, such conditions should insure consistency. For kernels with characteristic exponent of 2, $\alpha = .2$ may be (asymptotically) optimal. For densities with high frequency wiggles, small values of p should be used.

A recent approach of Carmichael and Parzen [8] attacks the probability density estimation problem using techniques motivated by the estimation of spectral densities of second order stationary time series. On the basis of a random sample $\{x_1, x_2, \ldots, x_n\}$ from an unknown density with domain $[-\pi, \pi]$, the estimate is

(211) $$\hat{f}_p(x) = \frac{\hat{K}_p}{2\pi} \frac{1}{\left| \sum_{j=0}^{p} \hat{\alpha}_{jp} e^{ijx} \right|^2},$$

where the parameters are obtained from the Yule–Walker equations

(212) $$\begin{bmatrix} \hat{R}(0) & \cdots & \hat{R}(p-1) \\ \vdots & & \vdots \\ \hat{R}(1-p) & & \hat{R}(0) \end{bmatrix} \begin{bmatrix} \hat{\alpha}_{1p} \\ \vdots \\ \hat{\alpha}_{pp} \end{bmatrix} = - \begin{bmatrix} \hat{R}(-1) \\ \vdots \\ \hat{R}(-p) \end{bmatrix}$$

and

$$\sum_{j=0}^{p} \hat{\alpha}_{jp} \hat{R}(j) = \hat{K}_p, \qquad \alpha_{0p} = 1$$

with

$$\hat{R}(v) = \frac{1}{n} \sum_{j=1}^{n} \int_{-\pi}^{\pi} e^{ivx} \delta(x - x_j) \, dx.$$

Under very general conditions

(213) $$|\hat{f}_p(x) - f(x)| \to 0 \text{ a.e. in } [-\pi, \pi]$$

provided that $\lim_{n \to \infty} p^3/n = 0$.

A lengthy paper of Boneva, Kendall, and Stefanov [6] uses the usual histogram as a starting point for a smoothed estimator of an unknown probability density function f with cdf F. Let the domain of the "histospline" estimator $[a, b]$ be divided into k equal width intervals with

(214)
$$[a, b] = \bigcup_{j=0}^{k-1} [a_j, a_{j+1}],$$

where

$$a_0 = a$$
$$a_k = b$$

and

$$a_{j+1} - a_j = \frac{b - a}{k}.$$

Then the histospline estimator $\hat{F}_H \in C^{(2)}[a, b]$ is the minimizer of

(215)
$$\int_a^b (\hat{F}'')^2 \, dx$$

subject to the constraint

$$F_H(a_j) = \frac{\text{number of samples} \leq a_j}{n}.$$

The solution is a cubic spline [40]. The estimator of the unknown density is then

(216)
$$\hat{f}_H(x) = \hat{F}'_H(x).$$

Since the authors do not add the constraint $F'_H \geq 0$, their density estimator can go negative.

Schoenberg [41] has proposed a simple, less "wiggly" and nonnegative alternative to the histospline. To use this procedure, we first construct an ordinary histogram on the interval $[a, b]$. Thus, if we have k intervals of length h with endpoints $\{a = a_0, a_1, a_2, \ldots, a_k = b\}$, we have y_j for histogram height for the jth interval $I_j = [a_{j-1}, a_j)$, where $y_j = $ (number of observations in I_j)$/nh$. Let $A_j = \left(a_j - \frac{h}{2}, y_j\right)$ for $j = 1, 2, \ldots, k$. We also mark the extreme vertices $B_0 = (a, y_1)$ and $B_k = (b, y_k)$. These points are connected with the polygonal line $B_0 A_1 A_2 \cdots A_{k-1} A_k B_k$. The successive intersections $\{B_1, B_2, \ldots, B_{k-1}\}$ with the verticle segments of the histogram are marked. Clearly, $B_j = \left(a_j, \frac{y_j + y_{j+1}}{2}\right)$ for $j = 1, 2, \ldots, k - 1$. The point A_j is equidistant from the verticle lines $x = a_{j-1}$ and $x = a_j$. We finally draw parabolic arcs $\{B_{j-1}B_j\}$ joining B_{j-1} to B_j, such that the parabola at B_{j-1} has tangent $B_{j-1}A_j$ and at B_j the tangent B_jA_j. Schoenberg's *splinegram* $s(x)$ is defined to be the quadratic spline, having as its graph the union of these arcs. A similar approach is given by Lii and Rosenblatt [30].

An algorithm of Good [21] examined by Good and Gaskins [22, 23] serves as the motivation for Chapters 4 and 5 of this study. We have already noted that in the case where the functional form of an unknown density f is known and characterized by a parameter θ we can assume a prior density h on this parameter and obtain, on the basis of a random sample $\{x_1, x_2, \ldots, x_n\}$ and the prior density the posterior density of θ via

$$(217) \qquad g(\theta|x) = \frac{\prod\limits_{j=1}^{n} f(x_j|\theta)h(\theta)}{\int_{\tau \in \Theta} \prod\limits_{j=1}^{n} f(x_j|\tau)h(\tau)\, d\tau} = \prod\limits_{j=1}^{n} f(x_j|\theta)h^*(\theta, x).$$

It is then possible to use, for example, the posterior mode or posterior mean as an estimator for θ.

If we move to the case where we do not know the functional form of the unknown density, a Bayesian approach becomes more difficult [32]. To avoid these complications, Good suggests [21] a "Bayesian in mufti" algorithm. Let $\mathscr{F}$ be an appropriate space of probability densities. We seek to maximize in $\mathscr{F}$ the criterion functional

$$(218) \qquad w(f) = \sum_{j=1}^{n} \log f(x_j) - \Phi(f),$$

where

$$\Phi(f) \geq 0$$

and

$$\Phi(f_0) < \infty \qquad \text{where } f_0 \text{ is the true density.}$$

It is clear Good's approach is formally quasi-Bayesian, since

$$(219) \qquad \exp[w(f)] = \prod_{j=1}^{n} f(x_j) \exp[-\Phi(f)]$$

looks very similar to the expression (217).

We have the following argument from Good and Gaskins [22] to establish that their estimate $\hat{f}$ converges to the unknown density f_0 in the sense that for $a < b$

$$\int_a^b \hat{f}(x)\, dx \to \int_a^b f_0(x)\, dx.$$

For any $\hat{f}$

$$(220) \qquad w(f_0) - w(\hat{f}) = \sum_{j=1}^{n} \log\left[\frac{f_0(x_j)}{\hat{f}(x_j)}\right] - \Phi(f_0) + \Phi(f)$$

$$\geq \sum_{j=1}^{n} \log\left[\frac{f_0(x_j)}{\hat{f}(x_j)}\right] - \Phi(f_0).$$

Thus,

$$E[w(f_0) - w(\hat{f})] \geq n \int \log\left[\frac{f_0(x)}{\hat{f}(x)}\right] f_0(x)\, dx - \Phi(f_0).$$

Now, since the logarithm is strictly concave, we have

$$(221) \quad -\int \log\left[\frac{f_0(x)}{\hat{f}(x)}\right] f_0(x)\, dx = \int \log\left[\frac{\hat{f}(x)}{f_0(x)}\right] f_0(x)\, dx < \log \int \frac{\hat{f}(x)}{f_0(x)} f_0(x)\, dx$$

$$= 0.$$

Thus, since the first term on the right is strictly positive and the second does not involve n, sooner or later, it is argued,

$$n \int \log\left[\frac{f_0(x)}{\hat{f}(x)}\right] f_0(x)\, dx - \Phi(f_0)$$

will be strictly positive. Then using Chebyshev's inequality, we have that in probability $w(f_0) - w(\hat{f})$ will ultimately be positive.

However, $\hat{f}(x)$ is not a fixed function but (assuming it exists) is the maximizer of the criterion functional in (218). It could be more accurately denoted $\hat{f}(x|\mathbf{x}, \Phi)$. We cannot, then, conclude, using the above argument, that

$$(222) \qquad E[w(f_0) - w(\hat{f})] \geq \sum_{j=1}^{n} \underset{x}{E} \log\left[\frac{f_0(x_j)}{\hat{f}(x_j|\mathbf{x}, \Phi)}\right] - \Phi(f_0)$$

is positive for any n, however large. We shall give a consistency theorem for a maximum penalized likelihood estimator in Chapter 5.

In order to guarantee that $\hat{f}$ be nonnegative, Good and Gaskins substitute γ^2 for f in (218), solve for the maximizer $\hat{\gamma}$, then obtain the solution $\hat{f} = \hat{\gamma}^2$. Thus, Good and Gaskins propose as an "equivalent" problem the maximization of

$$(223) \qquad w(\gamma^2) = \sum_{j=1}^{n} \log \gamma^2(x_j) - \Phi(\gamma^2).$$

Assuming $\gamma \in L^2(-\infty, \infty)$, we may represent γ as a Hermite series:

$$(224) \qquad \gamma_\infty(x) = \sum_{m=0}^{\infty} \gamma_m \varphi_m(x),$$

where the γ_m are real
and

$$\varphi_m = e^{-x^2/2} H_m(x) 2^{-m/2} \pi^{-1/4} (m!)^{-1/2}$$

$$H_m(x) = (-1)^m e^{x^2} \frac{d^m}{dx^m} e^{-x^2}.$$

If $\gamma \in L^1(-\infty, \infty)$ and is of bounded variation in every finite interval, then the series converges to $\gamma(x)$ at all points of continuity of f.

To guarantee that $\int_{-\infty}^{\infty} \gamma_\infty^2(x)\, dx = 1$, the following condition is imposed:

$$(225) \qquad\qquad \sum_{m=0}^{\infty} \gamma_m^2 = 1.$$

As penalty function, Good and Gaskins use

$$(226) \qquad \Phi(\gamma) = 4\alpha \int_{-\infty}^{\infty} \gamma'^2(x)\, dx + \beta \int_{-\infty}^{\infty} \gamma''^2(x)\, dx.$$

Hence, it is necessary to assume that γ' and γ'' are also contained in $L^2(-\infty, \infty)$. Thus, for any r the criterion function to be maximized is

$$(227) \quad Q(\gamma) = \sum_{j=1}^{n} \log \gamma_r(x_j) - 4\alpha \int_{-\infty}^{\infty} (\gamma_r'(x))^2\, dx - \beta \int (\gamma_r''(x))^2\, dx,$$

where

$$\gamma_r(x) = \sum_{m=0}^{r} \gamma_m \varphi_m(x)$$

subject to

$$\sum_{m=0}^{r} \gamma_m^2 = 1.$$

Good and Gaskins use standard Lagrange multiplier techniques to attempt to find the maximizer of (227). This gives rise to the system of equations

$$\sum_{m=0}^{r} \gamma_m^2 = 1$$

$$(228) \quad 2 \sum_{j=1}^{n} \varphi_k(x_j) \left\{ \sum_{m=0}^{r} \gamma_m \varphi_m(x_j) \right\}^{-1} - 4\alpha[2k\gamma_k - \{(k+1)(k+n)\}^{1/2}\gamma_{k+2}$$

$$- \{(k-1)k\}^{1/2}\gamma_{k-2}] - \beta \Big[3k^2\gamma_k + 3k\gamma_k - (2k+3)\{(k+1)(k+2)\}^{1/2}\gamma_{k+2}$$

$$- (2k-1)\{(k-1)k\}^{1/2}\gamma_{k-2} + \frac{1}{2}\{(k+1)(k+2)(k+3)(k+4)\}^{1/2}\gamma_{k+4}$$

$$+ \frac{1}{2}\{(k-3)(k-2)(k-1)k\}^{1/2}\gamma_{k-4} \Big] - 2\lambda\gamma_k = 0; \qquad k = 0, 1, 2, \dots, r.$$

Practically, Good and Gaskins suggest using the approximation $\hat{\gamma}_R(x) \sim \hat{\gamma}_\infty(x)$ where R is a value beyond which the $\hat{\gamma}_j(j > R)$ are less than .001. There are a number of unanswered questions:

1) Under what conditions does the square of the solution to (223) solve (218)?
2) Under what conditions is the solution to (218) consistent?
3) Under what conditions does $\hat{\gamma}_m(x)$ have a limit as $m \to \infty$?
4) If $\hat{\gamma}_\infty(x)$ exists, under what conditions is it a solution to (223)?

Question (1) is addressed in Chapter 4. A modified version of question (2) is given in Chapter 5. The fact that in some small sample cases where we have been able to obtain closed form solutions—when the solution to (223) is the solution to (218)—these are at variance with the graphical representations of $\hat{f}$ given in [22] leads us to believe that questions (3) and (4) are nontrivial.

References

[1] Andrews, D. F., Bickel, P. J., Hampel, F. R., Huber, P. J., Rogers, W. H., and Tukey, J. W. (1972). *Robust Estimates of Location*. Princeton: Princeton University Press.

[2] Bennett, J. O., de Figueiredo, R. J. P., and Thompson, J. R. (1974). "Classification by means of B-spline potential functions with applications to remote sensing." *The Proceedings of the Sixth Southwestern Symposium on System Theory*, FA3.

[3] Bergstrom, Harald (1952). "On some expansions of stable distributions." *Arkiv for Matematik* 2: 375–78.

[4] Bickel, P. J., and Rosenblatt, M. (1973). "On some global measures of the deviations of density function estimates." *Annals of Statistics* 1: 1071–95.

[5] Bochner, Salomon (1960). *Harmonic Analysis and the Theory of Probability*, Berkeley and Los Angeles: University of California Press.

[6] Boneva, L. I., Kendall, D. G., and Stefanov, I. (1971). "Spine transformations: Three new diagnostic aids for the statistical data-analyst." *Journal of the Royal Statistical Society, Series B* 33: 1–70.

[7] Brunk, H. D. (1976). "Univariate density estimation by orthogonal series." Technical Report No. 51, Statistics Department, Oregon State University.

[8] Carmichael, Jean-Pierre (1976). "The Autoregressive Method: A Method of Approximating and Estimating Positive Functions." Ph. D. dissertation, State University of New York, Buffalo.

[9] Cramér, Harald (1928). "On the composition of elementary errors." *Skandinavisk Aktuarietidskrift* 11: 13–74, 141–80.

[10] _____ (1946). *Mathematical Methods of Statistics*. Princeton: Princeton University Press.

[11] Davis, K. B. (1975). "Mean square error properties of density estimates." *Annals of Statistics* 3: 1025–30.

[12] Epanechnikov, V. A. (1969). "Nonparametric estimates of a multivariate probability density." *Theory of Probability and Its Applications* 14: 153–58.

[13] Fama, Eugene F., and Roll, Richard (1968). "Some properties of symmetric stable distributions." *Journal of the American Statistical Association* 63: 817–36.

[14] _____ (1971). "Parameter estimates for symmetric stable distributions." *Journal of the American Statistical Association* 66: 331–38.

[15] Farrell, R. H. (1972). "On best obtainable asymptotic rates of convergence in estimates of a density function at a point." *Annals of Mathematical Statistics* 43: 170–80.

[16] Feller, William (1966). *An Introduction to Probability Theory and Its Applications*, II, New York: John Wiley and Sons.

[17] de Figueiredo, R. J. P. (1974). "Determination of optimal potential functions for density estimation and pattern classification." *Proceedings of the 1974 International Conference on Systems, Man, and Cybernetics* (C. C. White, ed.) IEEE Publication 74 CHO 908–4, 335–37.

[18] Fisher, R. A. (1937). "Professor Karl Pearson and the method of moments." *Annals of Eugenics* 7: 303–18.

[19] Galton, Francis (1879). "The geometric mean in vital and social statistics." *Proceedings of the Royal Society* 29: 365–66.

[20] ———— (1889). *Natural Inheritance*. London and New York: Macmillan and Company.

[21] Good, I. J. (1971). "A non-parametric roughness penalty for probability densities." *Nature* 229: 29–30.

[22] Good, I. J., and Gaskins, R. A. (1971). "Nonparametric roughness penalties for probability densities." *Biometrika* 58: 255–77.

[23] ———— (1972). "Global nonparametric estimation of probability densities. *Virginia Journal of Science* 23: 171–93.

[24] Graunt, John (1662). *Natural and Political Observations on the Bills of Mortality*.

[25] Johnson, N. L. (1949). "Systems of frequency curves generated by methods of translation." *Biometrika* 36: 149–76.

[26] Kazakos, D. (1975). "Optimal choice of the kernel function for the Parzen kernel-type density estimators." Submitted for publication.

[27] Kendall, Maurice G., and Stuart, Alan (1958). *The Advanced Theory of Statistics*, 1. New York: Hafner Publishing Company.

[28] Kim, Bock Ki, and Van Ryzin, J. (1974). "Uniform consistency of a histogram density estimator and model estimation." MRC Report 1494.

[29] Kronmal, R. A., and Tarter, M. E. (1968). "The estimation of probability densities and cumulatives by Fourier series methods." *Journal of the American Statistical Association* 63: 925–52.

[30] Lii, Keh-Shin, and Rosenblatt, M. (1975). "Asymptotic behavior of a spline estimate of a density function." *Computation and Mathematics with Applications* 1: 223–35.

[31] Lindeberg, W. (1922). "Eine neue Herleitung des Expenentialgesetzes in der Wahrscheinlichkeitsrechnung." *Mathematische Zeitschrift* 15: 211–26.

[32] de Montricher, G. M. (1973). *Nonparametric Bayesian Estimation of Probability Densities by Function Space Techniques*. Doctoral dissertation, Rice University, Houston, Texas.

[33] Nadaraya, E. A. (1965). "On nonparametric estimates of density functions and regression curves." *Theory of Probability and Its Applications* 10: 186–90.

[34] Ord, J. K. (1972). *Families of Frequency Distributions*. New York: Hafner Publishing Company.

[35] Parzen, E. (1962). "On estimation of a probability density function and mode." *Annals of Mathematical Statistics* 33: 1065–76.

[36] Pearson, Karl (1936). "Method of moments and method of maximum likelihood." *Biometrika* 28: 34–59.

[37] Rényi, Alfred (1970). *Foundations of Probability*. San Francisco: Holden Day.

[38] Rosenblatt, M. (1956). "Remarks on some nonparametric estimates of a density function." *Annals of Mathematical Statistics* 27: 832–35.

[39] ———— (1971). "Curve estimates." *Annals of Mathematical Statistics* 42: 1815–42.

[40] Schoenberg, I. J. (1946). "Contributions to the problem of approximation of equidistant data by analytical functions." *Quarterly of Applied Mathematics* 4: 45–99, 112–14.

[41] ———— (1972). "Notes on spline functions II on the smoothing of histograms." MRC Technical Report 1222.

[42] Schuster, Eugene F. (1970). "Note on the uniform convergence of density estimates." *Annals of Mathematical Statistics* 41: 1347–48.

[43] Scott, David W. (1976). *Nonparametric Probability Density Estimation by Optimization Theoretic Techniques*. Doctoral dissertation, Rice University, Houston, Texas.

[44] Scott, D. W., Tapia, R. A., Thompson, J. R. (1977) "Kernel density estimation revisited." *Nonlinear Analysis* 1: 339–72.

[45] Shapiro, J. S. (1969). *Smoothing and Approximation of Functions*. New York: Van Nostrand–Reinhold.

[46] Shenton, L. R. (1951). "Efficiency of the method of moments and the Gram-Charlier Type A distribution." *Biometrika* 38: 58–73.

[47] Tarter, M. E., and Kronmal, R. A. (1970). "On multivariate density estimates based on orthogonal expansions." *Annals of Mathematical Statistics* 41: 718–22.

[48] _____ (1976). "An introduction to the implementation and theory of nonparametric density estimation." *American Statistician* 30: 105–12.

[49] Tukey J. W. (1976). "Some recent developments in data analysis." Presented at 150th Meeting of the Institute of Mathematical Statistics.

[50] _____ (1977). *Exploratory Data Analysis*. Reading, Mass.: Addison–Wesley.

[51] Van Ryzin, J. (1969). "On strong consistency of density estimates." *Annals of Mathematical Statistics* 40: 1765–72.

[52] Wahba, Grace (1971). "A polynomial algorithm for density estimation." *Annals of Mathematical Statistics* 42: 1870–86.

[53] _____ (1975). "Optimal convergence properties of variable knot, kernel and orthogonal series methods for density estimation." *Annals of Statistics* 3: 15–29.

[54] Watson, Geoffrey S. (1969). "Density estimation by orthogonal series." *Annals of Mathematical Statistics* 40: 1496–98.

[55] Watson, G. S., and Leadbetter, M. R. (1963). "On the estimation of the probability density, I." *Annals of Mathematical Statistics* 34: 480–91.

[56] Wegman, Edward J. (1969). "A note on estimating a unimodal density." *Annals of Mathematical Statistics* 40: 1661–67.

[57] _____ (1970). "Maximum likelihood estimation of a unimodal density function." *Annals of Mathematical Statistics* 41: 457–71.

[58] _____ (1970). "Maximum likelihood estimation of a unimodal density, II." *Annals of Mathematical Statistics* 41: 2169–74.

[59] _____. "Maximum likelihood estimation of a probability density function." To appear in *Sankya, Series A*.

[60] Whittle, P. (1958). "On the smoothing of probability density functions." *Journal of the Royal Statistical Society* (B) 20: 334–43.

[61] Woodroofe, Michael (1970). "On choosing a delta-sequence." *Annals of Mathematical Statistics* 41: 1665–71.

3

Maximum Likelihood Density Estimation

3.1. Maximum Likelihood Estimators

In this chapter we return to the classical maximum likelihood estimation procedure discussed in Chapter 2. We establish general existence and uniqueness results for the finite dimensional case, show that several popular estimators are maximum likelihood estimators, and, finally, show that the infinite dimensional case is essentially meaningless. This latter fact serves to motivate the maximum penalized likelihood density estimation procedures that will be presented in Chapter 4.

A good part of the material presented in this chapter originated in a preliminary version of [1]. At that time it was felt that the results were probably known and consequently were not included in the published version. However, we have not been able to find the material in this generality in the literature.

Consider the interval (a, b). As in the previous chapters we are again interested in the problem of estimating the (unknown) probability density function $f \in L^1(a, b)$ (Lebesgue integrable on (a, b)) which gave rise to the random sample $x_1, \ldots, x_n \in (a, b)$.

By the *likelihood* that $v \in L^1(a, b)$ gave rise to the random sample $x_1, \ldots, x_n$ we mean

(1) $$L(v) = \prod_{i=1}^{n} v(x_i).$$

Let H be a manifold in $L^1(a, b)$, and consider the following constrained optimization problem

(2)   maximize $L(v)$; subject to

$$v \in H, \quad \int_a^b v(t)\, dt = 1 \quad \text{and} \quad v(t) \ge 0 \qquad \forall t \in (a, b).$$

By a *maximum likelihood estimate* based on the random sample $x_1, \ldots, x_n$ and corresponding to the manifold H, we mean any solution of problem (2).

We now restrict our attention to the special case when H is a finite dimensional subspace (linear manifold) of $L^1(a, b)$. In the following section we show that this special case includes several interesting examples. For $y \in R^n$ the notation $y \ge 0$ ($y > 0$) means $y_i \ge 0$ ($y_i > 0$), $i = 1, \ldots, n$. In this application $\langle\,,\,\rangle$ denotes the Euclidean inner product in R^n (see Example I.1). The following three propositions will be useful in our analysis.

Proposition 1.   Given the positive integers $q_1, \ldots, q_n$, define $f: R^n \to R$ by

$$f(y) = \prod_{i=1}^n y_i^{q_i}.$$

Also given $\alpha \in R^n$, such that $\alpha > 0$, define T by

$$T = \{y \in R^n : \langle \alpha, y \rangle = 1 \quad \text{and} \quad y \ge 0\}.$$

Then f has a unique maximizer in T which is given by y^*, where

$$y_i^* = \frac{q_i}{N \alpha_i} \quad \text{and} \quad N = \sum_{i=1}^n q_i.$$

Proof.   Clearly, T is compact and f is continuous; hence by Theorem I.3 there exists a maximizer which we shall denote by y^*. Now, if for some $1 \le i \le n$, we have $y_i^* = 0$, then $f(y^*) = 0$; however, $y = \left(\dfrac{1}{\alpha_1}, \ldots, \dfrac{1}{\alpha_n}\right) \in T$ and $f(y) > 0$, which would be a contradiction. It follows that y^* is such that $y^* > 0$. From Theorem I.8 there exists λ, such that

(3) $$\nabla f(y^*) = \lambda \alpha.$$

Taking the gradient of f and using (3) leads to

(4) $$q_i f(y^*) = \lambda \alpha_i y_i^*, \qquad i = 1, \ldots, n.$$

From (4) and the fact that $\langle \alpha, y^* \rangle = 1$ we have

(5) $$\lambda = \lambda \langle \alpha, y^* \rangle = f(y^*) \sum_{i=1}^n q_i = N f(y^*).$$

Substituting this value for λ in (4) establishes the proposition; since we have proved that $\nabla f(y) = \lambda \alpha$ has a unique solution. ∎

Proposition 2. Consider $f: R^n \to R$ defined by

$$(6) \qquad\qquad f(y) = \prod_{i=1}^{n} y_i.$$

Let T be any convex subset of $R_+^n = \{y \in R^n : y \geq 0\}$ which contains at least one element $y > 0$. Then f has at most one maximizer in T.

Proof. Since there is at least one element $y > 0$ in T any maximizer of f in T will not lie on the boundary of R_+^n. Therefore, maximizing f over T is equivalent to maximizing the log of f over T, which is in turn equivalent to minimizing the negative log over T. However, by part (ii) of Proposition I.16 it is easy to see that $-\log f(y)$ with $f(y)$ given by (6) is a strictly convex function on the interior of R_+^n. The proposition now follows from Theorem I.2. ∎

Proposition 3. Let $\Phi_1, \ldots, \Phi_n$ be n linearly independent members of $L^1(a, b)$ and let

$$(7) \quad T_+ = \left\{ \alpha \in R^n : \int_a^b \sum_{i=1}^{n} \alpha_i \Phi_i(t)\, dt = 1 \quad \text{and} \quad \sum_{i=1}^{n} \alpha_i \Phi_i(t) \geq 0 \,\, \forall t \in [a, b] \right\}.$$

Then T_+ is a convex compact subset of R^n.

Proof. If T_+ is empty, we are through, so suppose that $\alpha^* = (\alpha_1^*, \ldots, \alpha_n^*) \in T_+$. Clearly, T_+ is closed and convex. Suppose T_+ is not bounded. Then there exists γ_m such that $(1 - \lambda)\alpha^* + \lambda\gamma_m \in T_+$ for $0 \leq \lambda \leq m$. Let $\beta_m = (1 - \lambda_m)\alpha^* + \lambda_m\gamma_m$, where $\lambda_m = \|\gamma_m - \alpha^*\|^{-1}$. Observe that $\|\beta_m - \alpha^*\| = 1$. Let β be any limit point of β_m. Then,

$$(8) \qquad\qquad (1 - \lambda)\alpha^* + \lambda\beta \in T_+, \qquad \forall \lambda \in [0, \infty).$$

Let $\delta = \beta - \alpha^*$. It follows from (8) that

$$\lambda \sum_{i=1}^{n} \delta_i \Phi_i(t) \geq -\sum_{i=1}^{n} \alpha_i^* \Phi_i(t), \qquad \forall \lambda \in [0, \infty);$$

hence,

$$\sum_{i=1}^{n} \delta_i \Phi_i(t) \geq 0 \qquad \forall t \in [a, b].$$

However,

$$\int_a^b \sum_{i=1}^{n} \delta_i \Phi_i(t)\, dt = 0, \qquad \text{so} \quad \sum_{i=1}^{n} \delta_i \Phi_i(t) = 0.$$

This contradicts the linear independence of $\Phi_1, \ldots, \Phi_n$ and proves the proposition. ■

We are now ready to prove our existence theorem.

Theorem 1 (Existence.) If H is a finite dimensional subspace of $L^1(a, b)$, then a maximum likelihood estimate based on $x_1, \ldots, x_n$ and corresponding to H exists.

Proof. Proposition 3 shows that in this case the constraint set of problem (2) is a convex compact subset of the finite dimensional space H. The existence of a solution of problem (2) now follows from Theorem I.3. ■

In addition to Theorem 1, we have the following weak form of uniqueness.

Theorem 2 (Uniqueness.) Suppose H is a finite dimensional subspace of $L^1(a, b)$, with the property that there exist at least one $\varphi \in H$ satisfying $\varphi(t) \geq 0 \ \forall t \in [a, b]$ and $\varphi(x_i) > 0, i = 1, \ldots, n$ for the random sample $x_1, \ldots, x_n$. If φ_1 and φ_2 are maximum likelihood estimates based on $x_1, \ldots, x_n$ and corresponding to H, then

(9) $$\varphi_1(x_i) = \varphi_2(x_i), \qquad i = 1, \ldots, n,$$

i.e., any two estimates must agree at the sample points.

Proof. The proof is a straightforward application of Proposition 2. ■

3.2. The Histogram as a Maximum Likelihood Estimator

In Chapter 2 we presented some results concerning the classical histogram and approaches for constructing generalized histograms. Let us return to the standard histogram. Consider a partition of the interval (a, b), say, $a = t_1 < t_2 < \cdots < t_{m+1} = b$. Let T_i denote the half-open half-closed interval $[t_i, t_{i+1})$ for $i = 1, \ldots, m$. Let $I(T_i)$ denote the indicator function of the interval T_i, i.e., $I(T_i)(x) = 0$, if $x \notin T_i$ and $I(T_i)(x) = 1$, if $x \in T_i$. For the random sample $x_1, \ldots, x_n \in (a, b)$, we let $M(T_i)$ denote the number of these samples which fall in the interval T_i. Clearly, $\sum_{i=1}^{m} M(t_i) = n$. Now, the classical theory (see, for example, Chapter 2 of Freund [2]) tells us that if we wish to construct a histogram with class intervals T_i, we make the heights of the rectangle with base T_i proportional to $M(T_i)$. This means that the histogram will have

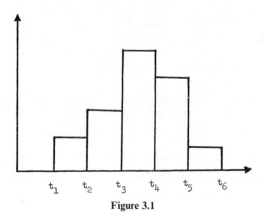

Figure 3.1

the form

(10)
$$f^* = \sum_{i=1}^{m} \frac{\alpha M(T_i)}{(t_{i+1} - t_i)} I(T_i)$$

for some constant of proportionality α. Since the area under the step function f^* must be equal to one, we must have

(11)
$$\sum_{i=1}^{m} \alpha M(T_i) = 1.$$

From (11) it follows that $\alpha = \dfrac{1}{n}$ and consequently the histogram estimator is given by

(12)
$$f^* = \sum_{i=1}^{m} \frac{M(T_i)}{n(t_{i+1} - t_i)} I(T_i).$$

The graph of f^* would resemble the graph in Figure 3.1.

In Chapter 2 we proved that the histogram estimator is actually a maximum likelihood estimator. For the sake of unity, and also to demonstrate the usefulness of Theorem 1, we will show that this result is actually a straightforward consequence of Theorem 1.

Proposition 4. The histogram (12) is the unique maximum likelihood estimate based on the random sample $x_1, \ldots, x_n$ and corresponding to the subspace of $L^1(a, b)$ defined by $S_0(t_1, \ldots, t_m) = \left\{ \sum_{i=1}^{m} y_i I(T_i) : y_i \in R \right\}$.

Proof. By Theorem 1 a maximum likelihood estimate

(13)
$$v^* = \sum_{i=1}^{m} y_i^* I(T_i)$$

exists. Observe that if $M(T_i) = 0$, then y_i^* in (13) is also equal to zero. For if $y_i^* > 0$, then we could modify v^* in (13) by setting $y_i^* = 0$ and increasing some $y_j^*(j \neq i)$, so that the constraints of problem (2) would still be satisfied, but the likelihood functional would be increased. However, this would lead to a contradiction, since v^* maximizes the likelihood over this constraint set. We therefore lose no generality by assuming $M(T_i) > 0 \; \forall i$. Let $q_i = M(T_i)$ and $\alpha_i = t_{i+1} - t_i$ for $i = 1, \ldots, m$. The constraints of problem (2) take the form $y \geq 0$ and $\langle \alpha, y \rangle = 1$ for $y_1 I(T_1) + \cdots + y_m I(T_m) \in S_0(t_1, \ldots, t_m)$. Proposition 1 now applies and says that (12) and (13) are the same. The uniqueness follows from Theorem 2, since the conditions are satisfied by this choice of H and (9) implies two maximum likelihood estimates would coincide on all intervals which contain sample points. And, as before, they would both be zero on intervals which contain no sample points.

Let $\{X_1, \ldots, X_m\}$ where $X_i < X_{i+1}$ denote the distinct samples in $\{x_1, \ldots, x_n\}$. Also, for each X_i, $i = 1, \ldots, m$, let q_i denote the number of samples equal to X_i so that $\sum_{i=1}^{m} q_i = n$. Finally, let X_0, X_{m+1} be two real numbers such that $a \leq X_0 < X_1$ and $X_m < X_{m+1} \leq b$.

Proposition 5. The maximum likelihood estimate based on $x_1, \ldots, x_n \in (a, b)$ and corresponding to the linear manifold

$$S_0(X_1, \ldots, X_m) = \left\{ \sum_{i=1}^{m} y_i I([X_i, X_{i+1})) : y_i \in R \right\} \subset L^1(a, b)$$

exists and is uniquely given by

$$(14) \qquad v^* = \sum_{i=1}^{m} y_i^* I([X_i, X_{i+1})),$$

where

$$(15) \qquad y_i^* = \frac{q_i}{n(X_{i+1} - X_i)}.$$

Proof. The proof is similar to the proof of Proposition 4 and follows from Proposition 1 by setting $\alpha_i = X_{i+1} - X_i$. ∎

Let $S_1(X_0, \ldots, X_{m+1})$ denote the subspace of $L^1(a, b)$ consisting of continuous functions which are linear on $[X_i, X_{i+1}]$, $i = 0, \ldots, m$ and vanish outside the interval (X_0, X_{m+1}).

Proposition 6. The maximum likelihood estimate based on $x_1, \ldots, x_n \in (a, b)$ and corresponding to the subspace $S_1(X_0, \ldots, X_{m+1})$ exists and is uniquely given by that v^* which satisfies

$$(16) \qquad v^*(X_i) = \frac{2q_i}{n(X_{i+1} - X_{i-1})}, \qquad i = 1, \ldots, m.$$

Proof. If we associate $v \in S_1(X_0, \ldots, X_{m+1})$ with $y \in R^m$, where $y_i = v(X_i)$, $i = 1, \ldots, m$, then we may again consider solving problem (2) by Proposition 1. Toward this end, let $\alpha_i = \frac{1}{2}(X_{i+1} - X_{i-1})$. The constraints of problem (2) become $y \geq 0$ and $\langle \alpha, y \rangle = 1$. The result now follows from Proposition 1 in the same way that Propositions 4 and 5 did. Again we have uniqueness because two members of $S_1(X_0, \ldots, X_{m+1})$ which agree at $X_0, \ldots, X_{m+1}$ must coincide. ∎

The notation $S_i(t_1, \ldots, t_m)$, $i = 1, 2$, used above actually denotes the class of polynomial splines of degree i with knots at the points $t_1, \ldots, t_m$. For example, when $i = 0$ we are working with functions which are piecewise constant, and for $i = 1$ we are working with continuous functions which are piecewise linear. It follows that the maximum likelihood estimate given by Proposition 5 is a histogram and resembles the histogram in Figure 3.1. Moreover, the maximum likelihood estimate given by Proposition 6 can be thought of as a generalized histogram and its graph would resemble the graph given in Figure 3.2.

The notion of the histogram as a polynomial spline of degree 0 could be extended not only to splines of degree 1 as in Proposition 6 but to splines of arbitrary degree, using either the proportional area approach or the maximum likelihood approach. However, in general, such an approach would create severe problems, including the facts that we could not obtain

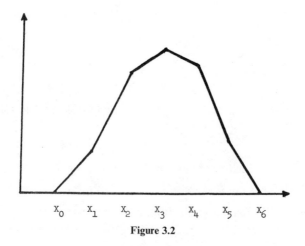

Figure 3.2

the estimate in closed form and it would be essentially impossible to enforce the nonnegativity constraint. For splines of degree 0 or 1 the nonnegativity constraint is a trivial matter, since they cannot "wiggle."

3.3. The Infinite Dimensional Case

In general, if the manifold H in problem (2) is infinite dimensional, then a maximum likelihood estimate will not exist. To see this observe that a solution is idealized by

$$(17) \qquad f^*(x) = \frac{1}{n} \sum_i \delta(x - x_i),$$

where δ is the Dirac delta mass at the origin. The linear combination of Dirac spikes f^* satisfies the constraints of problem (2) and gives a value of $+\infty$ for the likelihood. These comments are formal, since δ does not exist as a member of $L^1(a, b)$. However, they allow us to conclude that in any infinite dimensional manifold $H \subset L^1(a, b)$, which has the property that it is possible to approximate Dirac delta spikes, e.g., given $t^* \in (a, b)$ there exists $f_m \in H$, such that f_m integrates to one $f_m(t) \geq 0$ $\forall t \in (a, b)$ and $\lim_{m \to \infty} f_m(t^*) = +\infty$, the likelihood will be unbounded and maximum likelihood estimates will not exist. Moreover, most infinite dimensional manifolds in $L^1(a, b)$ will have this property. This is certainly true of the continuous functions, the differentiable functions, the infinitely differentiable functions, and the polynomials. Of course, we could consider pathological examples of infinite dimensional manifolds in $L^1(a, b)$, for which a maximum likelihood estimate exists. As an extreme case, let H be all continuous functions on (a, b) which vanish at the samples $x_1, \ldots, x_n$. In this case, any member of H, which is nonnegative and integrates to one is a maximum likelihood estimate based on $x_1, \ldots, x_n$.

The fact that in general the maximum likelihood estimate does not exist for infinite dimensional H has a very important interpretation in terms of finite dimensional H. Specifically, for large (but finite) dimensional H the maximum likelihood estimate must necessarily lead to unsmooth (contains spikes) estimates and a numerically ill-posed problem. Granted, if it is known a priori that the unknown probability density function is a member of the finite dimensional manifold H, then the maximum likelihood approach would probably give satisfactory results. However, this prior knowledge is known only in very special cases.

The above observations lead us to the following criticism of maximum likelihood estimation in general: for small dimensional H there is no flexibility and the solution will be greatly influenced by the subjective choice of H; while for large dimensional H the solution must necessarily be unsmooth and the optimization problem must necessarily create numerical problems. For example, notice that for a fixed random sample $x_1, \ldots, x_n$ the histogram estimate f^* given by (12) has the unfortunate property that $f^*(x) \to +\infty$ for $x \in \{x_1, \ldots, x_n\}$ and $f^*(x) \to 0$ for $x \notin \{x_1, \ldots, x_n\}$ as the number of intervals T_i goes to infinity and the length of T_i goes to zero. Whether or not a reasonable estimate is obtained is completely dependent on the delicate art of choosing T_i, $i = 1, \ldots, m$ properly, i.e., the interplay between the intervals T_i and the random sample $x_1, \ldots, x_n$. Indeed, this is a fundamental part of the consistency proof given in Section 2.5.

For the generalized histogram estimators given in Proposition 5 and Proposition 6 we are not free to choose the dimension of the problem (i.e., the linear manifold H) independent of the data, $x_1, \ldots, x_n$. Moreover, whether the estimate develops spikes as the sample size n becomes large, depends on the quantity $n \min_i (X_{i+1} - X_i)$, respectively, $n \min_i (X_{i+1} - X_{i-1})$, which the user has no control over.

In numerical analysis it is standard to handle an infinite dimensional problem by restricting the problem to a finite dimensional subspace (e.g., discrete mesh) and then using the solution of the approximate problem as an approximation to the solution of the original problem. However, this approach is usually predicated on the security that the infinite dimensional problem is well-posed and that as the dimension of the finite dimensional subspace becomes infinite (e.g., mesh size approaches zero) the approximate problem approaches the original problem in some reasonable manner. It is then a straightforward matter to establish convergence of the approximate solution to the true solution. In general, this is the situation with numerical methods for the solution of differential and integral equations. However, in many statistical applications the cart is put before the horse and considerable effort is spent in showing that the finite dimensional problem is well-posed, however, the asymptotic infinite dimensional problem is either ignored, meaningless or ill-posed. This latter phenomenon we choose to call *dimensional instability*. We further feel that dimensional stability is distinct from the numerical analysts' use of stability and the statisticians' use of robustness and that in some sense it implies these latter two notions. This notion probably merits further investigation.

In this section we have argued that the maximum likelihood density estimation problem as posed in Section 1 of this chapter is dimensionally unstable. The purpose of Chapter 4 is to describe one general approach for introducing dimensional stability into the maximum likelihood density estimation problem. In Chapter 5 we show how the theory developed in Chapter 4 can be used to develop a numerically efficient algorithm.

References

[1] de Montricher, G. M., Tapia, R. A., and Thompson, J. R. (1975). "Nonparametric maximum likelihood estimation of probability densities by penalty function methods." *Annals of Statistics* 3:1329–48.

[2] Freund, J. E. (1962). *Mathematical Statistics*. Englewood Cliffs, New Jersey: Prentice–Hall.

4

Maximum
Penalized Likelihood
Density Estimation

4.1. Maximum Penalized Likelihood Estimators

The material in this chapter is taken primarily from de Montricher, Tapia, and Thompson [1] and relies heavily on the theory of mathematical optimization in Hilbert space. A fairly complete (for the purposes of this chapter) treatment of mathematical optimization theory can be found in Appendix I. The reader not familiar with the theory will benefit by reading this appendix before embarking on the present material.

As in Chapter 3, let H be a manifold in $L^1(a, b)$ and consider a functional $\Phi : H \to R$. Given the random sample $x_1, \ldots, x_n \in (a, b)$ the Φ-*penalized likelihood* of $v \in H$ is defined by

$$(1) \qquad \hat{L}(v) = \prod_{i=1}^{n} v(x_i) \exp(-\Phi(v)).$$

Consider the constrained optimization problem:

(2) maximize $\hat{L}(v)$; subject to

$$v \in H, \qquad \int_a^b v(t)\, dt = 1 \qquad \text{and} \qquad v(t) \geqslant 0, \qquad \forall t \in (a, b).$$

The general form of the penalized likelihood (1) is due to Good and Gaskins [2]. Their specific suggestions are analyzed in Sections 3 and 4.

Any solution to problem (2) is said to be a *maximum penalized likelihood estimate* based on the random sample $x_1, \ldots, x_n$ corresponding to the manifold H and the penalty function Φ. Motivated by the observations made in Chapter 3 we will be particularly interested in the case when H is an infinite dimensional Hilbert space (see Appendix I.1). In the case that H

is a Hilbert space, a very natural penalty function to use is $\Phi(v) = \|v\|^2$ where $\|\circ\|$ denotes the norm on H. Consequently, when H is a Hilbert space and we refer to the penalized likelihood functional on H or to the maximum penalized likelihood estimate corresponding to H with no reference to the penalty functional Φ, we are assuming that Φ is the square of the norm in H. The Hilbert space inner product will be denoted by $\langle \circ, \circ \rangle$ so that $\langle x, x \rangle = \|x\|^2$.

For problem (2) to make sense we would like H to have the property that for $x_1, \ldots, x_n \in (a, b)$ there exists at least one $v \in H$, such that

$$(3) \qquad \int_a^b v(t)\, dt = 1, \qquad v(t) \geq 0 \qquad \forall t \in (a, b)$$

and

$$v(x_i) > 0 \qquad i = 1, \ldots, n.$$

Proposition 1. Suppose that H is a reproducing kernel Hilbert space and D is a closed convex subset of $\{v \in H : v(x_i) \geq 0, i = 1, \ldots, n\}$ with the property that D contains at least one function which is positive at the samples $x_1, \ldots, x_n$. Then the penalized likelihood functional (1) has a unique maximizer in D.

Proof. The $\langle \circ, \circ \rangle$-penalized likelihood functional $\hat{L}$ is clearly continuous when $\langle \circ, \circ \rangle$ represents the inner product in a reproducing kernel Hilbert space. By assumption there exists at least one $v \in D$, such that $\hat{L}(v) > 0$. Hence, maximizing $\hat{L}$ over D is equivalent to minimizing $J = -\log \hat{L}$ over D. A straightforward calculation gives the second derivative of J as

$$(4) \qquad J''(v)(\eta, \eta) = \sum_{i=1}^n \frac{\eta(x_i)\eta(x_i)}{v(x_i)^2} + 2\langle \eta, \eta \rangle.$$

The proposition now follows from Theorem 7 in Appendix I.4. ∎

Theorem 1. Suppose H is a reproducing kernel Hilbert space, integration over (a, b) is a continuous functional and there exists at least one $v \in H$ satisfying (3). Then the maximum penalized likelihood estimate corresponding to H exists and is unique.

Proof. The proof follows from Proposition 1, since the constraints in (2) give a closed convex subset of $\{v \in H : v(x_i) \geq 0, i = 1, \ldots, n\}$. ∎

The nonnegativity constraint in problem (2) is, in general, impossible to enforce when working with algorithms which deal with continuous densities which are not piecewise linear. For this reason, and others, we often find examples in the statistical literature where a problem with a nonnegativity constraint is solved by working with an equivalent problem stated in terms of the square root of the unknown density. In this latter problem, the non-negativity constraint is redundant. Specifically, given $H \subset L^1(a, b)$ and $J : H \to R$ consider the following two problems:

(5)   maximize $J(v)$; subject to

$$v \in H, \qquad \int_a^b v(t) \, dt = 1 \qquad \text{and} \qquad v(t) \geq 0 \qquad \forall t \in (a, b),$$

and

(6)   maximize $J(u^2)$; subject to

$$u^2 \in H \qquad \text{and} \qquad \int_a^b u^2(t) \, dt = 1.$$

The following proposition is obvious.

Proposition 2.

  (i) If v^* solves problem (5), then $u^* = \sqrt{v^*}$ solves problem (6).
  (ii) If u^* solves problem (6), then $v^* = (u^*)^2$ solves problem (5).

Part of the price one pays for no longer having to work with the non-negativity constraint is that the integral constraint is now nonlinear. In the following, when we consider the Good and Gaskins maximum penalized likelihood estimators, we will be dealing with constrained optimization problems of the form

(7)   maximize $J(v)$; subject to

$$\sqrt{v} \in \hat{H}, \qquad \int_a^b v(t) = 1 \qquad \text{and} \qquad v(t) \geq 0 \qquad \forall t \in (a, b),$$

where J is defined on $H \subset L^1(a, b)$ and $\hat{H}$ has the property that $w \in \hat{H}$ implies $w^2 \in H$. In order to avoid the nonnegativity constraint in problem (7), Good and Gaskins [2] suggested that one work with the analogous version of problem (6); namely,

(8)   maximize $J(u^2)$; subject to

$$u \in \hat{H} \qquad \text{and} \qquad \int_a^b u^2(t) \, dt = 1.$$

However, in this case there is a very subtle distinction between problems (7) and (8), and they are not always equivalent. Specifically, we have the following relationship between these two problems.

Proposition 3.

  If u^* solves problem (8), then $v^* = (u^*)^2$ solves problem (7) if and only if we have the additional condition that $|u^*| \in \hat{H}$.

That the additional condition in Proposition 3 is needed is clear from the requirement in problem (7) that $\sqrt{v} \in \hat{H}$. In the sequel, $\hat{H}$ will be a Hilbert space. Clearly, the condition that $u^* \in \hat{H}$ implies $|u^*| \in \hat{H}$ is trivially satisfied when $\hat{H} = R^n$. Moreover, if u^* is a nonnegative function, then this condition is also trivially satisfied. However, if u^* is a member of the Sobolev space $H^s(-\infty, \infty)$ (see Example 4 of Appendix I.1) or the restricted Sobolev space $H_0^s(a, b)$ (see Example 5 of Appendix I.1) which has a simple zero, then $|u^*|$ will also be a member of the space if and only if $s \leq 1$. Consequently, if we are working with $H^s(-\infty, \infty)$ or $H_0^s(a, b)$ with $s > 1$, and we can show that in our particular application problem (8) has a solution which takes on both positive and negative values, then the Good and Gaskins transformation is not valid. We will show in Section 4.4 that this is exactly the case with one of the estimators proposed by Good and Gaskins.

In Section 4.2 we will analyze the maximum penalized likelihood estimators proposed by de Montricher, Tapia, and Thompson [1]. The two estimators proposed by Good and Gaskins in [2] will be analyzed in Sections 4.3 and 4.4. The existence, uniqueness, and characterization results presented for these two estimators were established in [1] and were not considered by Good and Gaskins in [2].

Before going on to Section 4.2 we introduce a mathematical tool which will be extremely useful in the following three sections. For $f, g \in H_0^1(a, b)$ (see Appendix I.1) expressions of the form $\int_a^b f^{(j)}(t)g^{(k)}(t) \, dt$ make sense if $0 \leq j$, $k \leq 1$. Moreover, the integration by parts $\int_a^b f^{(j)}(t)g'(t) \, dt = -\int_a^b f^{(j+1)}(t)g(t) \, dt$ makes sense if and only if $j = 0$. However, by considering the $L^2(a, b)$ function f' as a distribution on (a, b) (see Horvath [3]) and interpreting derivatives as distributional derivatives we can write

$$\int_a^b f'(t)g'(t) \, dt = -\int_a^b f''(t)g(t) \, dt,$$

as long as we remember that f'' is a distribution and not a member of $L^2(a, b)$. This formal manipulation (which can be rigorously justified) offers us a very useful tool. For example, since $H_0^1(a, b)$ is a reproducing kernel Hilbert space (see Proposition 10 of Appendix I.2), we have by the Riesz representation theorem (see Theorem 1 of Appendix I.1) that there exists a unique $v_0 \in H_0^1(a, b)$, such that $v(0) = \int_a^b v_0'(t)v'(t) \, dt$ for all $v \in H_0^1(a, b)$ (we have assumed that $0 \in (a, b)$). If we proceed formally to integrate by parts, we obtain $v(0) = \int_a^b v_0'(t)v'(t) \, dt = -\int_a^b v_0''(t)v(t) \, dt$, and it follows that as a distribution $-v_0''$ is equal to δ the Dirac delta distribution. Recalling that δ is the distributional derivative of the distribution associated with $H(t)$, the Heaviside unit function ($H(t) = 0$ if $t \leq 0$ and $H(t) = 1$ if $t > 0$), it follows that $v_0'(t) = -H(t) + C_0$ and $v_0(t) = -S(t) + C_0 t + C_1$, where $S(t)$ is the continuous linear spline $S(t) = 0$, if $t \le 0$ and $S(t) = t$, if $t \ge 0$. Now, choosing the constants C_0 and C_1, so that $v_0(a) = v_0(b) = 0$, we see that $v_0 \in H_0^1(a, b)$ and v_0 is the Riesz representer of the evaluation functional $v \to v(0)$.

The above remarks clearly generalize to $H_0^s(a, b)$ and to $H^s(-\infty, \infty)$ (see Example 5 or Appendix I.1).

4.2. The de Montricher–Tapia–Thompson Estimator

Consider the restricted Sobolev space given in Example 5 of Appendix I.1; namely,

$$H_0^s(a, b) = \{f : f^{(j)} \in L^2(a, b), j = 0, \ldots, s \quad \text{and}$$
$$f^{(j)}(a) = f^{(j)}(b) = 0, j = 0, \ldots, s - 1\},$$

with inner product

(9)
$$\langle f, g \rangle = \int_a^b f^{(s)}(t) g^{(s)}(t) \, dt.$$

Theorem 2. The maximum penalized likelihood estimate corresponding to the Hilbert space $H_0^s(a, b)$ exists and is unique. Moreover, if the estimate is positive in the interior of an interval, then in this interval it is a polynomial spline (monospline) of degree 2s and of continuity class $2s - 2$ with knots exactly at the sample points.

Proof. The existence and uniqueness follow from Theorem 1, since Proposition 10 of Appendix I.2 shows that $H_0^s(a, b)$ is a reproducing kernel Hilbert space, Proposition 11 of Appendix I.2 shows that integration over the interval $[a, b]$ gives a linear functional which is continuous on $H_0^s(a, b)$ and there obviously exist members of $H_0^s(a, b)$ which satisfy (3).

When no confusion can arise, we will delete the variable of integration in definite integrals. Consider an interval $I_+ = [\alpha, \beta] \subset [a, b]$. Let $I_- = \{t \in [a, b] : t \notin [\alpha, \beta]\}$. Define the two functionals J_+ and J_- on $H_0^s(a, b)$ by

$$J_+(v) = -\sum_i \log v(x_i) + \int_{I_+} [v^{(s)}]^2,$$

and

$$J_-(v) = -\sum_i \log v(x_i) + \int_{I_-} [v^{(s)}]^2,$$

where the summation in the first formula is taken over all i, such that $x_i \in I_+$, and the summation in the second formula is taken over all i, such that $x_i \in I_-$. It should be clear that

$$J(v) = J_+(v) + J_-(v),$$

where as before $J(v) = -\log \hat{L}(v)$ and $\hat{L}$ is the penalized likelihood in $H_0^s(a, b)$. Let v_* denote the maximum penalized likelihood estimate for the samples $x_1, \ldots, x_n$. Suppose v_* is positive on the interval I_+. We claim that v_* solves the following constrained optimization problem:

(10)   minimize $J_+(v)$; subject to

$$v \in H_0^s(a, b), \; v^{(m)}(\alpha) = v_*^{(m)}(\alpha), \; v^{(m)}(\beta) = v_*^{(m)}(\beta), \qquad m = 0, \ldots, s-1,$$

$$\int_{I_+} v = \int_{I_+} v_* \quad \text{and} \quad v(t) \geq 0, \qquad t \in I_+.$$

To see this, observe that if v_+ satisfies the constraints of problem (10) and $J_+(v_*) > J_+(v_+)$, then the function v^* defined by

$$v^*(t) \begin{cases} = v_+(t), & t \in I_+ \\ = v_*(t), & t \in I_- \end{cases}$$

satisfies the constraints of problem (2), with $H_0^s(a, b)$ playing the role of H and $J(v_*) = J_+(v_*) + J_-(v_*) > J_+(v_+) + J_-(v_*) = J(v^*)$, which in turn implies that $\hat{L}(v_*) < \hat{L}(v^*)$; however, this contradicts the optimality of v_* with respect to problem (2). Now, define the functional G on $H_0^s(\alpha, \beta)$ by

$$G(v) = J(v_* + v) \quad \text{for} \quad v \in H_0^s(\alpha, \beta).$$

Consider the constrained optimization problem

(11)   minimize $G(v)$; subject to

$$v \in H_0^s(\alpha, \beta) \quad \text{and} \quad \int_{I_+} v = 0.$$

Suppose $v \neq 0$ is a solution of problem (11). Then, $(1-t)v_* + t(v_* + v) = v_* + tv$ satisfies the constraints of problem (10) for $t > 0$ and sufficiently small. By the strict convexity of J (see (4) and Appendix I.3) we have

(12) $$J(v_* + tv) < (1-t)J(v_*) + tJ(v_* + v) \leq J(v_*),$$

which contradicts the optimality of v_* with respect to problem (10). It follows that the zero function is the unique solution of problem (11). From the theory of Lagrange multipliers (see Appendix I.5), we must have

(13) $$\nabla G(0) + \lambda v_0 = 0,$$

where λ is a real number, $\nabla G(0)$ is the gradient of G at 0 (see Appendix I.3) and v_0 is the gradient of the functional $v \to \int_{I_+} v$ in the space $H_0^s(\alpha, \beta)$. Clearly, in this case v_0 is merely the Riesz representer of the functional $v \to \int_{I_+} v$ (see Appendix I.1). Specifically,

(14) $$\int_{I_+} v_0^{(s)} v^{(s)} = \int_{I_+} v.$$

Integrating by parts (in the distribution sense) we see that $v_0^{(2s)} = (-1)^s$; hence v_0 is a polynomial of degree $2s$ in $[\alpha, \beta]$. A straightforward calculation shows that

$$(15) \qquad \nabla G(0) = -\left(\sum_i \frac{v_i}{v_*(x_i)} - 2v_* \right),$$

where the summation is taken over i, such that $x_i \in I_+$ and v_i is the Riesz representer of the functional $v \to v(x_i)$ in $H_0^s(\alpha, \beta)$, i.e.,

$$(16) \qquad \int_{I_+} v^{(s)} v_i^{(s)} = v(x_i).$$

As before integrating by parts (in the distribution sense) we see that $v_i^{(2s)} = (-1)^s \delta_i$ where δ_i is the Dirac distribution at the point x_i. It follows that v_i is a polynomial spline of degree $2s - 1$ and of continuity class $2s - 2$, with a knot exactly at the sample point x_i. From (15) and (16) we have that v_* restricted to the interval $[\alpha, \beta]$ is a polynomial spline of degree $2s$ and of continuity class $2s - 2$, with knots exactly at the sample points in $[\alpha, \beta]$. A simple continuity argument takes care of the case when v_* is only positive on the interior of $[\alpha, \beta]$. Schoenberg [4] defines a monospline to be the sum of a polynomial of degree $2s$ and a polynomial spline of degree $2s - 1$. This proves the theorem. ∎

In Chapter 5 we will consider numerical algorithms based on Theorem 2 with $H_0^s(a, b)$ replaced by a finite dimensional subspace of $H_0^s(a, b)$.

4.3. The First Estimator of Good and Gaskins

Motivated by information theoretic considerations Good and Gaskins [2] consider the maximum penalized likelihood estimate corresponding to the penalty function

$$(17) \qquad \Phi_1(v) = \alpha \int_{-\infty}^{\infty} \frac{v'(t)^2}{v(t)} \, dt \qquad (\alpha > 0).$$

They did not define the manifold H, but it is obvious from the constraints that must be satisfied and the fact that

$$(18) \qquad \frac{1}{4} \Phi_1(v) = \alpha \int_{-\infty}^{\infty} \left[\frac{dv^{1/2}}{dt} \right]^2 \, dt$$

what the underlying manifold H should be, namely, $v^{1/2} \in H^1(-\infty, \infty)$ where $H^1(-\infty, \infty)$ is the Sobolev space given in Example 4 of Appendix I.1, namely, $H^1(-\infty, \infty) = \{f : R \to R : f' \text{ exists almost everywhere and } f,$

$f' \in L^2(-\infty, \infty)\}$ with inner product

(19) $$\langle f, g \rangle = \int_{-\infty}^{\infty} f(t)g(t) \, dt + \int_{-\infty}^{\infty} f'(t)g'(t) \, dt.$$

The situation here is going to be very delicate, because it is possible to show that the integration functional is *not* continuous in $H^1(-\infty, \infty)$. In the present application, problem (2) takes the form

(20) maximize $L_1(v) = \prod\limits_{i=1}^{n} v(x_i) \exp(-\Phi_1(v))$; subject to

$$v^{1/2} \in H^1(-\infty, \infty), \int_{-\infty}^{\infty} v(t) \, dt = 1 \quad \text{and} \quad v(t) \geq 0 \quad \forall t \in (-\infty, \infty).$$

In an effort to avoid the nonnegativity constraint in problem (20), Good and Gaskins considered working with $v^{1/2}$ instead of v. Specifically, if we let $u = v^{1/2}$, then restating problem (20) in terms of u we obtain

(21) maximize $\prod\limits_{i=1}^{n} u(x_i)^2 \exp\left(-4\alpha \int_{-\infty}^{\infty} u'(t)^2 \, dt\right)$; subject to

$$u \in H^1(-\infty, \infty) \quad \text{and} \quad \int_{-\infty}^{\infty} u(t)^2 \, dt = 1.$$

Problem (21) is solved for u^*, and then $v^* = (u^*)^2$ is accepted as the solution to problem (20). This transformation was discussed in Section 4.1, and Proposition 3 shows that since we are working with $H^1(-\infty, \infty)$, it is valid.

Problem (21) cannot possibly have a unique solution. To see this, notice that if u^* is a solution, then so is $-u^*$. Adding the nonnegativity constraint to problem (21) and restating in the form obtained by taking the square root of the objective functional (since it is nonnegative) we arrive at the following constrained optimization problem:

(22) maximum $\hat{L}(v) = \prod\limits_{i=1}^{n} v(x_i) \exp(-\Phi(v))$; subject to

$$v \in H^1(-\infty, \infty), \quad \cdot \int_{-\infty}^{\infty} v(t)^2 \, dt = 1 \quad \text{and} \quad v(t) \geq 0, \quad \forall t \in (-\infty, \infty)$$

where

(23) $$\Phi(v) = 2\alpha \int_{-\infty}^{\infty} v'(t)^2 \, dt$$

and α is given in (17).

Proposition 4.
   (i) If v solves problem (20), then $v^{1/2}$ solves problem (21) and problem (22).
   (ii) If u solves problem (21), then $|u|$ solves problem (22) and u^2 solves problem (20).

(iii) If v solves problem (22), then v solves problem (21) and v^2 solves problem (20).

Proof. The proof follows from Proposition 3 and the fact that if $v \geq 0$, then

$$\Phi(v^{1/2}) = \tfrac{1}{2}\Phi_1(v) \tag{24}$$

and

$$\hat{L}_1(v) = \hat{L}(v^{1/2})^2. \quad \blacksquare \tag{25}$$

Corollary 1. If problem (22) has a unique solution, then problem (20) has a unique solution; and although problem (22) cannot have a unique solution, it will have solutions, and the square of any of these solutions will give the unique solution of problem (20).

The remainder of this section is dedicated to demonstrating that problem (22) has a unique solution which is a positive exponential spline with knots only at the sample points. The same will then be true of Good's and Gaskins's first maximum penalized likelihood estimate.

Along with problem (22) we will consider the constrained optimization problem obtained by only requiring nonnegativity at the sample points:

(26) maximize $\hat{L}(v)$; subject to

$$v \in H^1(-\infty, \infty), \quad \int_{-\infty}^{\infty} v(t)^2 \, dt = 1 \quad \text{and} \quad v(x_i) \geq 0, \quad i = 1, \ldots, n.$$

Given $\lambda > 0$ and α in problem (22), we may also consider the constrained optimization problem:

(27) maximize $\hat{L}_\lambda(v) = \prod_{i=1}^{n} v(x_i) \exp(-\Phi_\lambda(v))$; subject to

$$v \in H^1(-\infty, \infty), \quad \int_{-\infty}^{\infty} v(t)^2 \, dt = 1 \quad \text{and} \quad v(x_i) \geq 0, \quad i = 1, \ldots, n,$$

where

$$\Phi_\lambda(v) = 2\alpha \int_{-\infty}^{\infty} v'(t)^2 \, dt + \lambda \int_{-\infty}^{\infty} v(t)^2 \, dt. \tag{28}$$

Our study of problem (27) will begin with the study of the following constrained optimization problem:

(29) maximize $\hat{L}_\lambda(v)$; subject to

$$v \in H^1(-\infty, \infty) \quad \text{and} \quad v(x_i) \geq 0, \quad i = 1, \ldots, n,$$

where $\hat{L}_\lambda$ is given by problem (27). Let $L^2 = L^2(-\infty, \infty)$.

Proposition 5. Problem (29) has a unique solution. Moreover, if v_λ denotes this solution, then

   (i) v_λ is an exponential spline with knots at the sample points $x_1, \ldots, x_n$;
   (ii) $v_\lambda(t) > 0, \forall t \in (-\infty, \infty)$; and
   (iii) $\|v_\lambda\|_{L^2} \geq (n/4\lambda))^{1/2}$.

Proof. From Proposition 12 of Appendix I.2 $H^1(-\infty, \infty)$ is a reproducing kernel space. Also $\|v\|_\lambda^2 = \Phi_\lambda(v)$ gives a norm equivalent to the original norm on $H^1(-\infty, \infty)$. The existence of v_λ now follows from Proposition 1 with $D = \{v \in H^1(-\infty, \infty) : v(x_i) \geq 0, i = 1, \ldots, n\}$. We will denote the Φ_λ inner product by $\langle \ , \ \rangle_\lambda$. Let v_i be the representer in the Φ_λ inner product of the continuous linear functional given by point evaluation at the point x_i, $i = 1, \ldots, n$, i.e.,

$$(30) \qquad \langle v_i, \eta \rangle_\lambda = \eta(x_i), \qquad \forall \eta \in H^1(-\infty, \infty).$$

Equivalently,

$$(31) \quad 2\alpha \int_{-\infty}^{\infty} v_i'(t)\eta'(t)\, dt + \lambda \int_{-\infty}^{\infty} v_i(t)\eta(t)\, dt = \eta(x_i), \qquad \forall \eta \in H^1(-\infty, \infty).$$

Integrating by parts (in the distribution sense) gives

$$(32) \quad \int_{-\infty}^{\infty} [-2\alpha v_i''(t) + \lambda v_i(t)]\eta(t)\, dt = \eta(x_i), \qquad \forall \eta \in H^1(-\infty, \infty);$$

hence,

$$(33) \qquad -2\alpha v_i'' + \lambda v_i = \delta_i, \qquad i = 1, \ldots, n$$

where $\delta_i(t) = \delta_0(t - x_i)$ and δ_0 denotes the Dirac distribution, i.e., $\int_{-\infty}^{\infty} \delta_0(t)\eta(t)\, dt = \eta(0)$. If we let v_0 be the solution of (33) for $i = 0$, then

$$(34) \qquad v_0(t) = \frac{1}{2(2\alpha\lambda)^{1/2}} \exp((\lambda(2\alpha))^{1/2}t), \qquad t < 0$$

$$= \frac{1}{2(2\alpha\lambda)^{1/2}} \exp(-(\lambda/(2\alpha))^{1/2}t), \qquad t > 0$$

and $v_i(t) = v_0(t - x_i)$ for $i = 1, \ldots, n$. Since v_λ is the maximizer we have that $v_\lambda(x_i) > 0, i = 1, \ldots, n$. We necessarily have that the derivative of $\hat{L}_\lambda$ at v_λ must be the zero functional; equivalently the gradient of $\hat{L}_\lambda$, or for that matter the gradient of $\log \hat{L}_\lambda$, must vanish at v_λ, since $\hat{L}_\lambda$ and $\log \hat{L}_\lambda$ have the same maxima. A calculation similar to that used in the proof of Proposition 1 gives

$$(35) \qquad \nabla_\lambda \log \hat{L}_\lambda(v) = 2v - \sum_{i=1}^{n} \frac{v_i}{v(x_i)},$$

where ∇_λ denotes the gradient. It follows from (35) that

$$(36) \qquad v_\lambda = \frac{1}{2} \sum_{i=1}^{n} \frac{v_i}{v_\lambda(x_i)}.$$

Properties (i) and (ii) are now immediate. Since $\langle v_i, v_\lambda \rangle_\lambda = v_\lambda(x_i)$, from (36) we have

$$(37) \qquad \|v_\lambda\|_\lambda^2 = n/2.$$

A straightforward calculation shows that

$$(38) \qquad v_i'(t)v_j(t) \leq \frac{\lambda}{2\alpha} v_i(t)v_j(t), \qquad \text{for} \quad i, j = 1, \ldots, n.$$

So

$$(39) \qquad v'(t)^2 = \frac{1}{4}\left[\sum_i \left(\frac{v_i'(t)}{v_\lambda(x_i)}\right)^2 + \sum_{i,j} \frac{v_i'(t)v_j'(t)}{v_\lambda(x_i)v_\lambda(x_j)}\right]$$

$$\leq \frac{\lambda}{8\alpha}\left[\sum_i \left(\frac{v_i(t)}{v_\lambda(x_i)}\right)^2 + \sum_{i,j} \frac{v_i(t)v_j(t)}{v_\lambda(x_i)v_\lambda(x_j)}\right] = \frac{\lambda}{2\alpha} v_\lambda(t)^2.$$

Integrating in t gives

$$(40) \qquad 2\alpha\|v_\lambda'\|_{L^2(-\infty, \infty)}^2 \leq \lambda\|v_\lambda\|_{L^2(-\infty, \infty)}^2.$$

By definition of the Φ_λ-norm and (37) we have property (iii). This proves the proposition. ∎

Proposition 6.   Problem (26) has a unique solution.

Proof.   Let $B = \{v \in H^1(-\infty, \infty): \int_{-\infty}^{\infty} v(t)^2 \, dt \leq 1 \text{ and } v(x_i) \geq 0, i = 1, \ldots, n\}$. Clearly, B is closed and convex. If $\hat{L}_\lambda$ is given by (27), then by Proposition 1 it will have a unique maximizer in B; say μ_λ. Now, by property (iii) of Proposition 5, if we choose $0 < \lambda < \frac{1}{4}$, then, v_λ, the unique solution of problem (29), will be such that $\|v_\lambda\|_{L^2(-\infty, \infty)} > 1$. We will show that for this range of λ, $\|\mu_\lambda\|_{L^2(-\infty, \infty)} = 1$. Consider $v_\theta = \theta v_\lambda + (1 - \theta)\mu_\lambda$. We know that $-\log \hat{L}_\lambda$ is a strictly convex functional (see the proof of Proposition 1). Moreover, $\log \hat{L}_\lambda(v_\lambda) \geq \log \hat{L}_\lambda(\mu_\lambda)$; hence $\log \hat{L}_\lambda(v_\theta) \geq \log \hat{L}_\lambda(\mu_\lambda)$ for $0 < \theta < 1$. Now suppose $\|\mu_\lambda\|_{L^2(-\infty, \infty)} < 1$ and consider

$$(41) \qquad g(\theta) = \|v_\theta\|_{L^2(-\infty, \infty)}.$$

We have $g(0) < 1$ and $g(1) > 1$. So for some $0 < \theta_0 < 1$, $g(\theta_0) = 1$ and $\log \hat{L}_\lambda(\mu_\lambda) \leq \log \hat{L}_\lambda(v_{\theta_0})$. This is a contradiction, since μ_λ is the unique maximizer of $\hat{L}_\lambda$ in B; hence, $\|\mu_\lambda\|_{L^2(-\infty, \infty)} = 1$. This shows that μ_λ is the unique solution of problem (27) for $0 < \lambda < \frac{1}{4}$. However, the term $\lambda \int_{-\infty}^{\infty} v(t)^2 \, dt$ is constant over the constraint set in problems (26) and (27); hence, problems (26) and (27) have the same solutions for any $\lambda > 0$. This proves the proposition, since we have demonstrated that problem (24) has a unique solution for at least one λ. ∎

Proposition 7. Problem (22) has a unique solution which is positive and an exponential spline with knots at the points $x_1, \ldots, x_n$.

Proof. If we can demonstrate that $\tilde{v}$, the unique solution of problem (26), has these properties we will be through. Let $G(v) = \log \hat{L}(v)$ where $\hat{L}$ is given in problem (22) and let

$$(42) \qquad g(v) = \int_{-\infty}^{\infty} v(t)^2 \, dt$$

for $v \in H^1(-\infty, \infty)$. Clearly, $\tilde{v}(x_i) > 0$ for $i = 1, \ldots, n$; hence, from the theory of Lagrange multipliers there exist λ, such that $\tilde{v}$ satisfies the equations

$$(43) \qquad G'(v) - \lambda g'(v) = 0 \qquad \text{and} \qquad g(v) = 1.$$

Using $L^2(-\infty, \infty)$ gradients (in the sense of distributions) (43) is equivalent to

$$(44) \qquad -4\alpha v'' + 2\lambda v = \sum_{i=1}^{n} \frac{\delta_i}{v(x_i)} \qquad \text{and} \qquad g(v) = 1,$$

where δ_i is the Dirac distribution at the point x_i. Since we have already established that problem (26) has a unique solution, it follows that (44) must have a unique solution in $H^1(-\infty, \infty)$; namely, $\tilde{v}$. If $\lambda \leq 0$, then any solution of the first equation in (44) would be a sum of trigonometric functions and could not possibly satisfy the constraint $g(v) = 1$, i.e., cannot be contained in $L^2(-\infty, \infty)$. It follows that $\lambda > 0$. Now observe that

$$G - \lambda g = \log \hat{L}_\lambda,$$

where $\hat{L}_\lambda$ is given by problem (27); hence, if $\tilde{v}$ satisfies (43) (from the first equation alone) it must also be a solution of problem (29) for this λ and therefore has the desired properties according to Proposition 5. This proves the proposition. ∎

Theorem 3. The first maximum penalized likelihood estimate of Good and Gaskins exists and is unique; specifically, the maximum penalized likelihood estimate corresponding to the penalty function

$$\Phi(v) = \alpha \int_{-\infty}^{\infty} \frac{v'(t)^2}{v(t)} \, dt \qquad (\alpha > 0)$$

and the manifold

$$H = \{v : v \geq 0 \quad \text{and} \quad v^{1/2} \in H^1(-\infty, \infty)\}$$

exists and is unique. Moreover, the estimate is positive and an exponential spline with knots only at the sample points.

Proof. The proof follows from Proposition 4 and Proposition 7. ∎

4.4. The Second Estimator of Good and Gaskins

Consider the functional $\Phi : H^2(-\infty, \infty) \to R$ defined by

$$(45) \qquad \Phi(v) = \alpha \int_{-\infty}^{\infty} v'(t)^2 \, dt + \beta \int_{-\infty}^{\infty} v''(t)^2 \, dt$$

for some $\alpha \geq 0$ and $\beta > 0$. By a *second maximum penalized likelihood estimate of Good and Gaskins* we mean any solution of the following constrained optimization problem:

(46)   maximize $\hat{L}_1(v) = \prod_{i=1}^{n} v(x_i) \exp(-\Phi(v^{1/2}))$; subject to

$$v^{1/2} \in H^2(-\infty, \infty), \quad \int_{-\infty}^{\infty} v(t) \, dt = 1 \quad \text{and} \quad v(t) \geq 0 \quad \forall t \in (-\infty, \infty).$$

As in the first case (described in the previous section), Good and Gaskins suggest avoiding the nonnegativity constraint by calculating the solution of problem (46) from the following constrained optimization problem:

(47)   maximize $\prod_{i=1}^{n} v(x_i)^2 \exp(-\Phi(v))$; subject to

$$v \in H^2(-\infty, \infty) \quad \text{and} \quad \int_{-\infty}^{\infty} v(t)^2 \, dt = 1,$$

where Φ is given by (45).

Along with problem (47) we consider the constrained optimization problem:

(48)   maximize $\hat{L}(v) = \prod_{i=1}^{n} v(x_i) \exp(-\tfrac{1}{2}\Phi(v))$; subject to

$$v \in H^2(-\infty, \infty), \int_{-\infty}^{\infty} v(t)^2 \, dt = 1 \quad \text{and} \quad v(x_i) \geq 0, \quad i = 1, \ldots, n.$$

Problem (48) was obtained from problem (47) by taking the square root of the functional to be maximized (since it is nonnegative) and requiring nonnegativity at the sample points; hence, the two problems only differ by the nonnegativity constraints at the sample points. This simple difference will allow us to establish uniqueness of the solution of problem (48); whereas problem (47) cannot have a unique solution. We will presently demonstrate that the solutions of problem (47) and problem (48) are not necessarily nonnegative. It will then follow that we cannot obtain the solution of problem (46) by considering problem (47) and problem (48). If we naively use v_*^2, where v_* solves problem (48), as an estimate for the probability density function giving rise to the random sample $x_1, \ldots, x_n$, then, clearly, v_*^2 will be nonnegative and integrate to one and is therefore a probability density; however, the estimate obtained in this manner will not in the strict sense of our definition be the maximum penalized likelihood estimate corresponding to problem (46), i.e., the second maximum penalized likelihood estimate of Good and Gaskins. For this reason we will refer to this latter estimate as the *pseudo maximum penalized likelihood estimate of Good and Gaskins.*

The next six propositions are needed to show that the second maximum penalized likelihood estimate and the pseudo maximum penalized likelihood estimate of Good and Gaskins exist, are unique, and are distinct.

Given $\lambda > 0$ consider the constrained optimization problem:

$$(49) \quad \text{maximize } \hat{L}_\lambda(v) = \prod_{i=1}^{n} v(x_i) \exp(-\Phi_\lambda(v)); \text{ subject to}$$

$$v \in H^2(-\infty, \infty), \quad \int_{-\infty}^{\infty} v(t)^2 \, dt = 1 \quad \text{and} \quad v(x_i) \geq 0, \quad i = 1, \ldots, n,$$

where

$$\Phi_\lambda(v) = \Phi(v) + \lambda \int_{-\infty}^{\infty} v(t)^2 \, dt$$

with $\Phi(v)$ given by (45).

As before we also consider the constrained optimization problem obtained by dropping the integral constraint:

$$(50) \quad \text{maximize } \hat{L}_\lambda(v); \text{ subject to}$$

$$v \in H^2(-\infty, \infty), \quad \text{and} \quad v(x_i) \geq 0, \quad i = 1, \ldots, n.$$

Proposition 8. Problem (49) has a unique solution. Moreover, if v_λ denotes this solution, then

$$\|v_\lambda\|_{L^2(-\infty, \infty)} \to +\infty \quad \text{as} \quad \lambda \to 0.$$

Proof. By Proposition 11 of Appendix I.2 the Sobolev space $H^2(-\infty, \infty)$ is a reproducing kernel Hilbert space. Moreover, if

$$\|v'\|_\lambda^2 = \Phi_\lambda(v),$$

then an integration by parts gives

$$(51) \qquad \|v'\|_{L^2}^2 = |\langle v, v''\rangle_{L^2}| \leq \|v\|_{L^2}\|v''\|_{L^2}$$
$$\leq \tfrac{1}{2}[\|v\|_{L^2}^2 + \|v''\|_{L^2}^2],$$

where L^2 denotes $L^2(-\infty, \infty)$; hence, $\|\circ\|_\lambda$ is equivalent to the original norm on $H^2(-\infty, \infty)$. The existence and uniqueness of v_λ now follows from Proposition 1.

We must now show that $\|v_\lambda\|_{L^2} \to +\infty$ as $\lambda \to 0$. From the fundamental theorem of calculus we have

$$(52) \qquad v(x)^2 = \int_{-\infty}^x \frac{dv(t)^2}{dt}\, dt = 2\int_{-\infty}^x v(t)v'(t)\, dt$$
$$\leq 2\|v\|_{L^2}\|v'\|_{L^2}.$$

Also, $\|v''\|_{L^2} \leq \|v\|_\lambda/\beta^{1/2}$, so that from (51) and (52)

$$(53) \qquad v(x)^2 \leq 2\|v\|_{L^2}^{3/2}(\|v\|_\lambda/\beta^{1/2})^{1/2}.$$

Evaluating (53) at x_i, taking logs (since $v(x_i) \geq 0$) and summing over i gives

$$(54) \qquad \sum_{i=1}^n \log v(x_i) \leq \frac{n}{4}\log\left(\frac{4}{\beta^{1/2}}\|v\|_\lambda\right) + \frac{3n}{4}\log(\|v\|_{L^2}).$$

Hence, from (54) we see that

$$(55) \qquad \log \hat{L}_\lambda(v) \leq \frac{3n}{4}\log(\|v\|_{L^2}) + \frac{n}{4}\log\left(\frac{4}{\beta^{1/2}}\|v\|_\lambda\right) - \|v\|_\lambda^2.$$

In a manner exactly the same as that used to establish (37), we have that $\|v_\lambda\|_\lambda^2 = n/2$. Hence, from (55) and the fact that $\log \hat{L}_\lambda(v) \leq \log \hat{L}_\lambda(v_\lambda)$ we obtain

$$(56) \qquad \log \hat{L}_\lambda(v) \leq \frac{3n}{4}\log(\|v_\lambda\|_{L^2}) + \frac{n}{8}\log(8n/\beta) - \frac{n}{2},$$

for any $v \in \{u \in H^2(-\infty, \infty): u(x_i) \geq 0, i = 1, \ldots, n\}$.

Let a and b be such that

$$a < \min_i(x_i) \qquad \text{and} \qquad \max_i(x_i) < b.$$

Given $\lambda > 0$ and ϵ and δ define the function θ_λ in the following piece-wise fashion:

$$\begin{aligned}
\theta_\lambda(t) &= \lambda^\epsilon \exp(-(t-a)^2/2\sigma^2) && \text{for} \quad t \in (-\infty, a) \\
&= \lambda^\epsilon && \text{for} \quad t \in [a, b] \\
&= \lambda^\epsilon \exp(-(t-b)^2/2\sigma^2) && \text{for} \quad t \in (b, +\infty),
\end{aligned}$$

where $\sigma = \lambda^\delta$. Straightforward calculations can be used to show

$$\log\left(\prod_{i=1}^n \theta_\lambda(x_i) \right) = \epsilon n \log(\lambda),$$

$$\|\theta_\lambda\|_{L_2}^2 = (b-a)\lambda^{2\epsilon} + ((\pi\lambda)^{1/2})^{2\epsilon-\delta},$$

$$\|\theta_\lambda'\|_{L_2}^2 = ((2\pi\lambda)^{1/2})^{2\epsilon-\delta},$$

$$\|\theta_\lambda''\|_{L_2}^2 = 2((2\pi\lambda)^{1/2})^{2\epsilon-3\delta},$$

and

$$(57) \quad \|\theta_\lambda\|_\lambda^2 = (b-a)\lambda^{2\epsilon+1} + ((\pi\lambda)^{1/2})^{2\epsilon+\delta+1} + 4\alpha((2\pi\lambda)^{1/2})^{2\epsilon-\delta}$$
$$+ 2\beta((2\pi\lambda)^{1/2})^{2\epsilon-3\delta}.$$

If we want $\|\theta_\lambda\|_\lambda^2 \to 0$ as $\lambda \to 0$, it is sufficient to choose all exponents of λ in (57) positive. If we also want

$$\log\left(\prod_{i=1}^n \theta_\lambda(x_i) \right) \to +\infty \qquad \text{as} \quad \lambda \to 0,$$

we should choose $\epsilon < 0$. This leads to the inequalities

$$(58) \qquad\qquad 2\epsilon + 1 > 0$$
$$2\epsilon + \delta + 1 > 0$$
$$2\epsilon - \delta > 0 \qquad\qquad ;$$
$$2\epsilon - 3\delta > 0$$
$$\epsilon < 0.$$

The system of inequalities (58) has solutions; specifically, $\epsilon = -\frac{1}{32}$ and $\delta = -\frac{1}{8}$ is one such solution. With this choice of ϵ and δ we see that $\log \hat{L}_\lambda(\theta_\lambda) \to +\infty$ as $\lambda \to 0$. It follows from (56) by choosing $v = \theta_\lambda$ that $\|v_\lambda\|_{L_2} \to +\infty$ as $\lambda \to 0$. This proves the proposition. ∎

Theorem 4. Problem (48) has a unique solution, i.e., the pseudo maximum penalized likelihood estimate of Good and Gaskins exists and is unique.

Proof. By Proposition 8 there exists $\lambda > 0$, such that if v_λ is the unique solution of problem (49), then $\|v_\lambda\|_{L_2} > 1$. Now, if $B = \{v \in H^2(-\infty, \infty): \int_{-\infty}^\infty v(t)^2 \, dt \le 1 \text{ and } v(x_i) \ge 0, \ i = 1, \ldots, n\}$, then B is closed and convex.

The proof of the theorem is now exactly the same as the proof of Proposition 6. ∎

By the change of unknown function $v \to v^{1/2}$ we see that problem (46) is equivalent to the following constrained optimization problem:

(59)   maximize $\hat{L}(v) = \prod_{i=1}^{n} v(x_i) \exp(-\tfrac{1}{2}\Phi(v))$; subject to

$$v \in H^2(-\infty, \infty), \quad \int_{-\infty}^{\infty} v(t)^2 \, dt = 1 \quad \text{and} \quad v(t) \geq 0 \quad \forall t \in (-\infty, \infty),$$

where $\Phi(v)$ is given by (45).

In turn, for $\lambda > 0$ problem (59) is equivalent to

(60)   maximize $\hat{L}_\lambda(v)$: subject to

$$v \in H^2(-\infty, \infty), \quad \int_{-\infty}^{\infty} v(t)^2 \, dt = 1 \quad \text{and} \quad v(t) \geq 0 \quad \forall t \in (-\infty, \infty),$$

where $\hat{L}_\lambda$ is defined in problem (49).

As in the previous two cases, we also consider the constrained optimization problem:

(61)   maximize $\hat{L}_\lambda(v)$; subject to

$$v \in H^2(-\infty, \infty) \quad \text{and} \quad v(t) \geq 0 \quad \forall t \in (-\infty, \infty),$$

where $\hat{L}_\lambda(v)$ is defined in problem (49).

Proposition 9. Problem (61) has a unique solution. Moreover, if v_λ^+ denotes this solution, then

$$\|v_\lambda^+\|_{L_2} \to +\infty \quad \text{as} \quad \lambda \to 0.$$

Proof. The existence of v_λ^+ follows from Proposition 1 as in the proof of Proposition 8. Let us first show that

(62) $$\|v_\lambda^+\|_\lambda \leq (n/2)^{1/2}.$$

Let $J_\lambda = -\log \hat{L}_\lambda$.

Clearly, for all nonnegative η in $H^2(-\infty, \infty)$ we have

(63) $$J_\lambda'(v_\lambda^+)(\eta - v_\lambda^+) \geq 0.$$

To see this, suppose that for some η (63) does not hold. Then by the definition of the Gâteaux derivative (see Appendix I.3) we will have

(64) $$J_\lambda(v_\lambda^+ + t(\eta - v_\lambda^+)) < J(v_\lambda^+)$$

for $t > 0$ sufficiently small. However, $v_\lambda^+ + t(\eta - v_\lambda^+) = (1 - t)v_\lambda^+ + t\eta \geq 0$

for $t > 0$ sufficiently small. Hence (64) contradicts the optimality of v_λ^+ with respect to problem (61). A straightforward calculation shows that

$$J_\lambda'(v)(\eta) = \sum_{i=1}^{n} \frac{\eta(x_i)}{v(x_i)} - 2\langle v, \eta \rangle_\lambda;$$

hence,

(65) $$J_\lambda'(v_\lambda^+)(v_\lambda^+) = n - 2\|v_\lambda\|_\lambda^2.$$

Now, choosing $\eta = 0$ in (63) and using (65) we arrive at (62). The functions θ_λ defined in the proof of Proposition 8 satisfy the constraints of this problem; hence,

(66) $$\log \hat{L}_\lambda(\theta_\lambda) \leq \log \hat{L}_\lambda(v_\lambda^+).$$

From (55) and (62) we have

(67) $$\log \hat{L}_\lambda(\theta_\lambda) \leq \frac{3n}{4} \log(\|v_\lambda^+\|_{L_2}) + \frac{n}{8} \log\left(\frac{8n}{\beta}\right) + \frac{n}{2}.$$

The proof now follows from (67), since $\log \hat{L}_\lambda(\theta_\lambda) \to +\infty$ as $\lambda \to 0$. ∎

Theorem 5. The second maximum penalized likelihood estimate of Good and Gaskins exists and is unique.

Proof. Using Proposition 9, the argument used to prove Theorem 4 shows that problem (60) has a unique solution, which is also the unique solution of problem (59). ∎

Theorem 6. The second maximum penalized likelihood estimate and the pseudo maximum likelihood estimate of Good and Gaskins are distinct.

Proof. We will show that it is possible for problem (48) to have solutions which are not nonnegative. Toward this end let $n = 1$, $x_1 = 0$, $\alpha = 0$, and $\beta = 2$. Let $G(v) = \log \hat{L}(v)$, i.e.,

$$G(v) = \log v(0) - \int_{-\infty}^{\infty} v''(t)^2 \, dt$$

and let

$$g(v) = \int_{-\infty}^{\infty} v(t)^2 \, dt.$$

As in the proof of Proposition 7, using the theory of Lagrange multipliers, we see that the solutions of problem (48) in this case are exactly the solutions of

(68) $$v^{(iv)} + \lambda v = \frac{\delta_1}{2v(0)} \quad \text{and} \quad g(v) = 1,$$

where δ_1 is defined in the proof of Proposition 7. If we let $\tilde{v}$ denote the Fourier transform of v, then taking the Fourier transform of the first expression in (68) gives

$$\tilde{v}(w) = [2v(0)(\lambda + 16\pi^4 w^4)]^{-1}.$$

Since $\|\tilde{v}\|_{L^2(-\infty, \infty)} = \|v\|_{L^2(-\infty, \infty)} = 1$ we must have

$$(69) \qquad \int_{-\infty}^{\infty} \frac{dw}{(\lambda + 16\pi^4 w^4)^2} = 4v(0)^2.$$

For the integral in (69) to exist we must have $\lambda > 0$. Now, the inverse Fourier transform of $(\lambda + 16\pi^4 w^4)^{-1}$ is given by v, where

$$(70) \qquad v(t) = \frac{e^{bt}}{8b^3}[\cos bt - \sin bt], \qquad t \leq 0$$

$$= \frac{e^{-bt}}{8b^3}(\cos bt + \sin bt], \qquad t > 0$$

with $b = \lambda^{1/4}/2^{1/2}$. From (70) $v(0) = (8b^3)^{-1}$ and from (69) $v(0)^2 = \frac{1}{4}\lambda^{-7/4K}$, where $K = \|(1 + 16\pi^4 w^4)^{-1}\|_{L^2(-\infty, \infty)}$. Hence, $\lambda^{1/4} = 2K$ and $b = 2^{1/2}K$. It follows that v is not nonnegative. Moreover, $|v|$ does not have a continuous derivative, so $|v| \notin H^2(-\infty, \infty)$. ∎

Corollary 1. The second maximum penalized likelihood estimate of Good and Gaskins cannot be obtained by solving problem (47).

Proof. Observe that the solution constructed in the proof of Theorem 6 is also a solution of problem (47) for this example. ∎

References

[1] de Montricher, G. M., Tapia, R. A., and Thompson, J. R. (1975). "Nonparametric maximum likelihood estimation of probability densities by penalty function methods." *Annals of Statistics* 3:1329–48.

[2] Good, I. J., and Gaskins, R. A. (1971). "Nonparametric roughness penalties for probability densities." *Biometrika* 58: 255–77.

[3] Horvath, J. (1966). *Topological Vector Spaces and Distributions*. Reading, Massachusetts: Addison-Wesley.

[4] Schoenberg, I. J. (1968). "Monosplines and quadrature formulae," in *Theory and Application of Spline Functions* (T.N.E. Greville, ed.). New York: Academic Press.

5

Discrete Maximum Penalized Likelihood Estimation

5.1. Discrete Maximum Penalized Likelihood Estimators

Our examination of penalized likelihood density estimations will now be brought to the state of practical computation. The arguments below are due mainly to Scott [5] and to Scott, Tapia, and Thompson [7]. As before, given the random sample $x_1, \ldots, x_n$ we would like to estimate, on an interval (a, b), the unknown density function f which gave rise to this random sample. From Chapter 4 our estimates are required to integrate to one on (a, b) and have support in (a, b); hence it follows that if the support of the unknown density f is not contained in (a, b), then we will actually be estimating the density function $\bar{f}$, which is close to f in the following sense

(1)
$$\bar{f}(x) = \begin{cases} \dfrac{f(x)}{\int_a^b f(x)\,dx} & \text{if } \quad a < x < b \\ 0 & \text{otherwise.} \end{cases}$$

Notice that f and $\bar{f}$ will coincide if and only if (a, b) contains the support of f. For the random sample $x_1, \ldots, x_n$ it is assumed that all observations outside (a, b) have been censored.

We will construct approximations to the maximum penalized likelihood estimate corresponding to the Hilbert space $H_0^1(a, b)$ (see Theorem 2 of Chapter 4) by solving finite dimensional versions of problem (2) in Chapter 4, which are reasonable approximations to the infinite dimensional problem. The terminology "discrete" is used because our finite dimensional versions of problem (4.2) will be obtained by working with the finite dimensional linear manifolds which arise by introducing a discrete mesh on

(a, b) and then considering the polynomial spline spaces of functions which are piece-wise constant or piece-wise linear on this mesh. Specifically, for a given positive integer m consider the uniform mesh $a = t_0 < t_1 < \cdots < t_m = b$, where $t_i = a + ih$, $i = 0, \ldots, m$, with $h = (b - a)/m$. Let T_i denote the half-open half-closed interval $[t_{i-1}, t_i)$, for $i = 1, \ldots, m$, and let $I(T_i)$ denote the indicator function of the interval T_i, i.e., $I(T_i)(x) = 0$, if $x \notin T_i$ and $I(T_i)(x) = 1$, if $x \in T_i$. Let $S_0(t_0, \ldots, t_m)$ denote the functions with support in $[t_0, t_m)$, equivalently $[a, b)$, and which are constant on the intervals T_i. Finally, let $S_1(t_0, \ldots, t_m)$ denote the continuous functions with support in (t_0, t_m) and which are linear on the intervals T_i. A typical member of $S_0(t_0, \ldots, t_m)$ would have a graph similar to that given by Figure 3.1 in Section 3.2 of Chapter 3, while a typical member of $S_1(t_0, \ldots, t_m)$ would have a graph similar to that given by Figure 3.2 in Section 3.2 of Chapter 3.

Since each $s_0 \in S_0(t_0, \ldots, t_m)$ has the representation

$$(2) \qquad s_0(x) = \sum_{i=1}^{m} y_{i-1} I(T_i)(x),$$

where

$$(3) \qquad y_i = s_0(t_i), \qquad i = 0, \ldots, m - 1,$$

it follows that $S_0(t_0, \ldots, t_m)$ is isomorphic to R^m. Moreover, assuming the correspondence (3), we see that

$$(4) \qquad s_0(x) \geq 0 \qquad \forall x \in (a, b) \Leftrightarrow y_i \geq 0, \qquad i = 0, \ldots, m - 1$$

and

$$(5) \qquad \int_a^b s_0(x)\, dx = 1 \Leftrightarrow h \sum_{i=0}^{m-1} y_i = 1.$$

Furthermore, since each $s_1 \in S_1(t_0, \ldots, t_m)$ has the representation

$$(6) \qquad s_1(x) = \sum_{i=1}^{m} [y_{i-1} + h^{-1}(x - t_{i-1})(y_i - y_{i-1})] I(T_i)(x),$$

where

$$(7) \qquad y_i = s_1(t_i), \qquad i = 1, \ldots, m - 1,$$

and by assumption $y_0 = s_1(t_0) = 0$ and $y_m = s_1(t_m) = 0$. It follows that $S_1(t_0, \ldots, t_m)$ is isomorphic to R^{m-1}. Moreover, assuming the correspondence (6), we see that

$$(8) \qquad s_1(x) \geq 0 \qquad \forall x \in (a, b) \Leftrightarrow y_i \geq 0, \qquad i = 1, \ldots, m - 1$$

and

(9)
$$\int_a^b s_1(x)\, dx = 1 \Leftrightarrow h \sum_{i=1}^{m-1} y_i = 1.$$

Let us now consider the $H_0^1(a, b)$ inner product defined on $S_i(t_0, \ldots, t_m)$, $i = 0, 1$, i.e., for s_0 given by (2) and s_1 given by (6) we are interested in the quantities

(10)
$$\|s_i\|_H^2 = \int_a^b s_i'(x)^2\, dx, \qquad i = 0, 1.$$

For $i = 1$, this creates no problem, since $S_1(t_0, \ldots, t_m) \subset H_0^1(a, b)$ and it follows that

(11)
$$\|s_1\|_s^2 = h^{-1} \sum_{i=1}^{m} (y_i - y_{i-1})^2.$$

However, since $S_0(t_0, \ldots, t_m) \not\subset H_0^1(a, b)$ the expression (10) is meaningless, but (11), even in this case, makes perfectly good sense, provided we interpret y_m to be zero. It therefore seems reasonable that if we wish to consider $S_i(t_0, \ldots, t_m)$, $i = 0, 1$ as approximations of the Hilbert space $H_0^1(a, b)$ we should work with the inner product

(12)
$$\langle s, \bar{s} \rangle_s = \alpha h^{-1} \sum_{i=1}^{m} (y_i - y_{i-1})(\bar{y}_i - \bar{y}_{i-1}) \qquad (\alpha > 0)$$

where $s, \bar{s} \in S_i(t_0, \ldots, t_m)$ and $y_i, \bar{y}_i$ are given by (2) and (3) or (6) and (7). Observe that (12) is actually a discrete Sobolev inner product approximating the continuous Sobolev inner product. Moreover, the constant α in (12) amounts to a scaling of the discrete Sobolev norm and will be important in our applications. We are now in a position to state our finite dimensional versions of problem (2) in Chapter 4. Specifically, given the random sample $x_1, \ldots, x_n$ consider the following constrained optimization problems.

(13)   Maximize $L_h^0(y_0, \ldots, y_{m-1}) = \prod_{i=1}^{n} s_0(x_i) \exp\left[-\alpha h^{-1} \sum_{i=1}^{m} (y_i - y_{i-1})^2 \right]$;

subject to

$$h \sum_{i=0}^{m-1} y_i = 1 \qquad \text{and} \qquad y_i \geq 0, \qquad i = 0, \ldots, m-1,$$

where s_0 is given by (2) and $y_m = 0$.

(14)   Maximize $L_h^1(y_1, \ldots, y_{m-1}) = \prod_{i=1}^{n} s_1(x_i) \exp\left[-\alpha h^{-1} \sum_{i=1}^{m} (y_i - y_{i-1})^2\right]$;

subject to

$$h \sum_{i=1}^{m} y_i = 1 \quad \text{and} \quad y_i \geq 0, \quad i = 1, \ldots, m-1$$

where s_1 is given by (6) and $y_0 = y_m = 0$.

The constant spline (member of $S_0(t_0, \ldots, t_m)$) corresponding to the solution of problem (13) and the linear spline (member of $S_1(t_0, \ldots, t_m)$) corresponding to the solution of problem (14) are called *discrete maximum penalized likelihood estimates* (DMPLE).

Theorem 1. The DMPLE exist and are unique.

Proof. It is a straightforward matter to show that (12) is actually an inner product on $S_i(t_0, \ldots, t_m)$, $i = 0, 1$. Moreover, since these inner product spaces are finite dimensional they must be reproducing kernel Hilbert spaces and integration must be continuous (see Appendix I). The theorem now follows from Theorem 1 of Chapter 4. ∎

5.2. Consistency Properties of the DMPLE

The following theorem is actually very satisfying, since it shows that, in contrast to the histogram described in Chapter 3, the DMPLE are dimensionally stable. From this point of view, we should think of DMPLE as "stable histograms." The proof of the theorem is quite lengthy and will not be given. The interested reader is referred to Scott, Tapia, and Thompson [7], where a detailed account can be found.

Theorem 2. Let h be the size of the mesh used to obtain the DMPLE guaranteed by Theorem 1. Then the constant spline DMPLE converges in sup norm as $h \to 0$ to the quadratic monospline MPLE guaranteed by Theorem 2 of Chapter 4 for $H_0^1(a, b)$. Furthermore, the linear spline DMPLE converges in $H_0^1(a, b)$ norm as $h \to 0$ to this same monospline MPLE.

We now turn to the very important question of consistency of the DMPLE. Let s_h denote the constant spline DMPLE given by problem (13) for a particular value of h. Recall that $\bar{f}$ is given by (1) and $x_1, \ldots, x_n$ is our random sample.

Theorem 3. Consider the constant spline DMPLE, with the number of mesh partitions given by $m = [Cn^q]$ (so $h = 0(n^{-q})$) where $C > 0$, $0 <$

$q < \frac{1}{4}$ and $[d]$ denotes the integer closest to d. Suppose that f is continuous on (a, b). Then for $x \epsilon (a, b)$

$$\lim_{n \to \infty} s_h(x) = \overline{f}(x) \text{ almost surely (a.s.).}$$

Proof.   For the sample $x_1, \ldots, x_n$, let

$$v_0 = \text{number of samples in } (-\infty, a)$$
$$v_k = \text{number of samples in } [t_{k-1}, t_k), \qquad k = 1, \ldots, m$$
$$v_{m+1} = \text{number of samples in } [b, \infty).$$

Then $\sum_{k=0}^{m+1} v_k = n$ and we define $n' = \sum_{k=1}^{m} v_k$. For this particular situation problem (13) is equivalent to

(15)   maximize $\log(L_h^0) = \sum_1^m v_k \log(y_k) - \alpha h^{-1} \sum_0^{m-1} (y_{k+1} - y_k)^2$

subject to

$$h \sum_0^{m-1} y_k = 1 \qquad \text{and} \qquad y_k \geq 0, \qquad k = 0, \ldots, m-1.$$

From the theory of Lagrange multipliers for equality and inequality constraints there exist multipliers $\lambda, \mu_0, \ldots, \mu_{m-1}$ such that for the solution of (15) we must have

(16) $\dfrac{\partial[\log(L_h^0)]}{\partial y_i} + \dfrac{\lambda \partial \left[h \sum_0^{m-1} y_k - 1 \right]}{\partial y_i} + \mu_i = 0, \qquad i = 0, \ldots, m-1,$

and the complementarity condition

(17) $$\mu_i y_i = 0, \qquad i = 0, \ldots, m-1.$$

Condition (16) becomes

(18) $$\mu_i + \frac{v_i}{y_i} + \frac{2\alpha}{h} \delta^2 y_i + \lambda = 0, \qquad i = 1, \ldots, m-1,$$

where

(19) $$\delta^2 y_i = y_{i+1} - 2y_i + y_{i-1}.$$

Multiplying (18) by y_i, summing over i, using (17), recalling the definition of n' and the first constraint in problem (15) we obtain the following expression for the multiplier λ

$$(20) \qquad \lambda = -n'h - 2\alpha \sum_{i=1}^{m-1} y_i\, \delta^2 y_i.$$

Substituting (20) into (18) and dividing by nh our necessary conditions become

$$(21) \qquad \frac{\mu_i}{nh} + \frac{v_i}{nhy_i} + \frac{2\alpha}{nh^2}\,\delta^2 y_i - \frac{n'}{n} - \frac{2\alpha}{nh}\sum_{j=1}^{m-1} y_j\,\delta^2 y_j = 0.$$

From the constraints of problem (15) we know that $0 \le y_j \le \dfrac{1}{h}$ so that

$$(22) \qquad \delta^2 y_j = 0\!\left(\frac{1}{h}\right)$$

and

$$(23) \qquad \sum_{j=1}^{m-1} y_j\,\delta^2 y_j = 0\!\left(\frac{1}{h^3}\right).$$

Using (22) and (23) in (21), we obtain for our necessary condition

$$(24) \qquad \frac{v_i}{nhy_i} - \frac{n'}{n} + 0\!\left(\frac{1}{nh^4}\right) = 0, \qquad i = 0, \ldots, m-1.$$

We are really interested in the interval containing x, and in this context the subscript i is misleading, since it depends on x and n. Consequently, let us introduce the notation $[x{:}n]$ for the mesh interval containing x for a particular value of n and let $v[x{:}n]$ denote the number of samples in $[x{:}n]$. Observe that $s_h(x) = y_i$ in this context, so that (24) implies

$$(25) \qquad \frac{v[x{:}n]}{nhs_h(x)} - \frac{n'}{n} + 0\!\left(\frac{1}{nh^4}\right) = 0.$$

The behavior of the first quantity in (25) is not at all obvious, and we will need the following lemma to complete the proof of the theorem.

Lemma 1.   Under the conditions of Theorem 3

$$\lim_{n\to\infty} \frac{v[x{:}n]}{nh} = f(x) \qquad \text{a.s.}$$

Proof.   We shall employ an argument motivated by Section 5.1 of [3]. In order to emphasize the dependence on n, we write $m(n)$ for the number of partitions. In this case the mesh spacing will be $h_n = (b - a)/m(n)$. Given the random sample $\{x_1, \ldots, x_n\}$, define the triangular array of random variables

$$Y_{nj} = \frac{I[x:n](x_j) - p[x:n]}{h_n},$$

where $I[x:n]$ denotes the indicator function of the interval $[x:n]$ and

$$p[x:n] = \int_{[x:n]} f(t) \, dt.$$

Now $\{Y_{nj}: j = 1, \ldots, n\}$ forms an independent sequence for each n, each random variable having zero mean. Also $I[x:n](x_j)$ is a binomial random variable with expectation given by $p[x:n]$. Let

$$S_n = \sum_{j=1}^{n} Y_{nj}.$$

To prove the lemma, we wish to show that $S_n/n \to 0$ almost surely. We have

$$\text{Var}(S_n) = np[x:n](1 - p[x:n])/h_n^2.$$

Hence, by Chebyshev's inequality for any $\epsilon > 0$,

(26) $$P\{|S_n| > n\epsilon\} < \frac{(1 - p[x:n])p[x:n]}{nh_n^2\epsilon^2}.$$

From the Borel–Cantelli lemma, a sufficient condition that S_n/n converge almost surely to zero is that

$$\sum_{n=1}^{\infty} P\{|S_n| > n\epsilon\} < +\infty.$$

From the Mean Value Theorem $p[x:n] = h_n f(\xi_n)$, where $\xi_n \in [x:n]$. Combining this fact with (26) leads to

(27) $$\sum_{n=1}^{\infty} P\{|S_n| > n\epsilon\} \le \sum_{n=1}^{\infty} \frac{[1 - h_n f(\xi_n)]f(\xi_n)}{nh_n\epsilon^2}.$$

Since the denominator of the summand on the right-hand side of (27) is $0(n^{1-q})$, this latter series diverges. However, if we consider the subsequence $\{n^2\}$, then $h_{n^2}n^2 = 0(n^{2-2q})$ with $0 < q < \frac{1}{4}$. Hence the majorizing series in (27) is convergent, and thus, by the Borel–Cantelli lemma

$$\frac{S_{n^2}}{n^2} \to 0 \qquad \text{a.s.}$$

Next, let

$$D_n = \max\{|S_k - S_{n^2}| : n^2 \le k < (n+1)^2\}; \quad n = 1, 2, \ldots.$$

We have for $n^2 \le k < (n+1)^2$

(28)
$$E[(S_k - S_{n^2})^2] = \sum_{j=1}^{n^2} (E[Y_{kj}^2] + E[Y_{n^2j}^2] - 2E[Y_{kj}Y_{n^2j}])$$

$$+ \sum_{j=n^2+1}^{k} E[Y_{kj}^2].$$

In order to deal with (28), we make the following important observations. A straightforward application of the Mean Value Theorem to the function $g(x) = x^{2q}$ shows that

$$Cn^{2q} - C(n+1)^{2q} = 0(n^{2q-1}).$$

However, since $m(n) = [Cn^q]$ is integer valued, we see that, with probability $0(n^{2q-1})$, $m(n^2) \ne m(k)$ for $n^2 \le k < (n+1)^2$ and it follows that, with probability $0(n^{2q-1})$, $[x:n^2] \ne [x:k]$ for k in this range.

A somewhat lengthy calculation using the Mean Value Theorem and the definitions shows that for $n^2 \le k < (n+1)^2$

$$E[Y_{kj^2}] = \frac{f(x)}{h_k} + 0(n^{2q}),$$

$$E[Y_{n^2j}^2] = \frac{f(x)}{h_{n^2}} + 0(n^{2q}),$$

and

$$E[Y_{kj}Y_{n^2}] = \frac{f(x)h_{kn^2}}{h_k h_{n^2}} + 0(n^{2q}),$$

where h_{kn^2} denotes the width of the interval $[x:k] \cap [x:n^2]$. Since $h_{n^2} \ne h_k$, with probability $0(n^{2q-1})$, it follows that (28) can be written

(29) $$E[(S_k - S_{n^2})^2] = \sum_{j=1}^{n^2} 0(n^{2q})0(n^{2q-1}) + \sum_{j=n^2+1}^{k} \left[\frac{f(x)}{h_k} + 0(n^{2q})\right] = 0(n^{4q+1}).$$

Now

$$E[D_n^2] \le \sum_{j=n^2}^{(n+1)^2} E[(S_k - S_{n^2})^2] = 0(n^{4q+2}).$$

So from Chebyshev's inequality, we have

$$P[D_n > n^2\epsilon] \le 0(n^{-2+2q}).$$

Hence, $\sum_{n=1}^{\infty} P[D_n > n^2\epsilon] < \infty$, since $0 < q < \frac{1}{4}$.

But, then by the Borel–Cantelli lemma

(30)
$$\frac{D^n}{n^2} \to 0 \qquad \text{a.s.}$$

Using the fact that for $n^2 \le k < (n+1)^2$

$$\left| \frac{S_k}{k} \right| \le \frac{|S_{n^2}| + D_n}{n^2},$$

we see that

$$\frac{S_n}{n} \equiv \frac{v[x:n]}{nh_n} - \frac{p[x:n]}{h_n} \to 0 \qquad \text{a.s.}$$

By the continuity of f we see that

$$\lim_{n \to \infty} \frac{p[x:n]}{h_n} = f(x),$$

so that, finally

(31)
$$\frac{v[x:n]}{nh_n} \to f(x) \qquad \text{a.s.}$$

This establishes the lemma. ∎

We now return to the proof of the theorem. The strong law of large numbers also implies that

(32)
$$\lim_{n \to \infty} \frac{n'}{n} = \int_a^b f(t)\, dt \qquad \text{a.s.}$$

Observing that $nh^4 \to \infty$ as $n \to \infty$, using (31) and (32) and letting $n \to \infty$ in (25) we establish the theorem. ∎

Remark. If in Theorem 3 we considered a discrete k-th order derivative in the penalty term of our criterion functional, then the analogous consistency result would require $0 < q < (2k+2)^{-1}$.

Remark. Numerical experience indicates that the requirement $0 < q < (2k+2)^{-1}$ is an artifact of the method of proof and is not needed for consistency.

5.3. Numerical Implementation and Monte Carlo Simulation

The numerical examples in this section were calculated using the linear spline DMPLE which is obtained from the penalized likelihood, using a discrete second derivative in the penalty term. We made this choice mainly because the linear spline gives a continuous estimate, and the use of the second derivative in the penalty term seems to give "fuller" estimates. Specifically, given a random sample $x_1, \ldots, x_n$, an interval (a, b) a positive scalar α and a positive integer m we let

$$(33) \qquad\qquad h = (b - a)/m,$$

$$(34) \qquad\qquad t_i = a + ih, \qquad i = 0, \ldots, m,$$

$$(35) \qquad\qquad p_0 = p_m = 0$$

and solve the $m - 1$ dimensional constrained optimization problem

$$(36) \quad \text{maximize } L(p_1, \ldots, p_{m-1}) = \sum_{i=1}^{n} \log p(x_i) - \frac{\alpha}{h} \sum_{k=1}^{m-1} [p_{k+1} - 2p_k + p_{k-1}]^2$$

subject to

$$h \sum_{k=1}^{m-1} p_k = 1 \qquad \text{and} \qquad p_k \geq 0, \qquad k = 1, \ldots, m - 1,$$

where

$$(37) \qquad p(t) = \begin{cases} p_k + \dfrac{p_{k+1} - p_k}{h} (t - t_k) & t \in [t_k, t_{k+1}) \\ 0 & t \notin (t_0, t_m). \end{cases}$$

A computer program has been written to solve problem (36) and has been incorporated in the IMSL program library [4]. This program uses the modification of Newton's method due to Tapia [9], [10] and is described in Appendix II.1. The nonnegativity constraints are handled as in Appendix II.2. This algorithm takes advantage of the special banded structure of the Hessian matrix of the Lagrangian functional for problem (36). Thus the amount of work turns out to be $O(m)$ instead of the expected $O(m^3)$ per iteration. Notice that the sample size n enters only in evaluating the gradient and Hessian and not in the matrix inversions. No initial guesses for Newton's method are required. A bootstrap algorithm is employed. First the problem is solved for $m = 7$, an easy problem to solve. This estimate provides initial guesses for $m = 11$, then for $m = 19$, and so on. The bootstrap algorithm takes advantage of the robustness of the DMPLE with respect to h.

The choice of α is very important and more difficult than the selection of the mesh spacing h. Asymptotically, of course, any positive value gives consistent results. For finite sample sizes; however, the choice is critical. The α parameter is analogous to the kernel scaling parameter h (for a discussion on kernel estimators see Section 2.5 of Chapter 2). Unfortunately (as is customary) we have also denoted the mesh spacing by h. Hopefully, this will not lead to confusion in the sequel where we will be discussing the robustness of the penalty parameter α and the kernel scaling parameter h. For a fixed sample $x_1, \ldots, x_n$ there are values of α and h which give the "best" approximations for the DMPLE and the kernel estimator respectively. For values smaller than "best," the corresponding estimates peak sharply at the sample points. On the other hand, values larger than "best" correspond to depressed and oversmoothed estimates. For the kernel estimator, asymptotically optimal choices for h can be derived; however, knowledge of the true (unknown) density is required. Scott, Tapia, and Thompson [8] have developed an iterative approach for estimating the optimal h. This approach was described in part in Section 2.5 of Chapter 2. Wahba [11] has also given considerable insight into the problem. In practice, an interactive approach is often used, where the smallest value of α (or h for the kernel estimator) is chosen that reveals fine structure without "too much" oscillatory behavior (consistent with prior knowledge). However, we note that the DMPLE is robust with respect to the choice of α for a choice of mesh. It is bounded from above and below and has finite support (the kernel estimator can approximate Dirac spikes and, on the other extreme, a very diffuse uniform density).

We next give several examples of density estimation problems which were solved using the DMPLE given by the solution of problem (36). Following these examples we present a Monte Carlo simulation to show that the DMPLE compares quite favorably with kernel estimators and is somewhat more robust.

Let us first consider the problem of estimating the bimodal density

$$(38) \quad f(x) = .75 \frac{1}{\sqrt{2\pi}} \exp\left[-\frac{1}{2}(x + 1.5)^2\right] + .25\left(\frac{3}{\sqrt{2\pi}} \exp\left[-\frac{9}{2}(x - 1.5)^2\right]\right).$$

In Figures 5.1 through 5.6, we show the effect of varying α over 5 $\log_{10}$ units for the DMPLE solution to problem (36) when $n = 300$. We note that discrete maximum penalized likelihood estimation lends itself particularly well to an evolutionary approach. We start with an oversmoothed α of 1,000 in Figure 5.1. Since the graph in Figure 5.2, using $\alpha = 100$, is significantly different, and no high frequency wiggles have yet appeared, we reduce α to 10 in Figure 5.3. Still no high frequency wiggles are in evidence. In Figure 5.4,

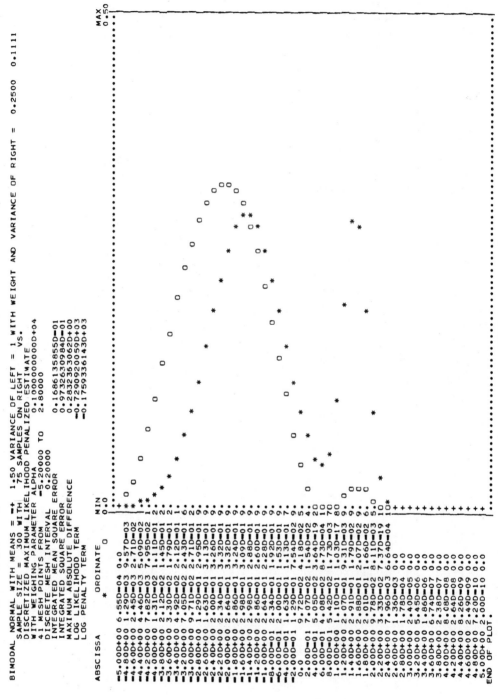

Figure 5.1. $n = 300$ Bimodal DMPLE $\alpha = 10^3$.

BIMODAL NORMAL WITH MEANS = -+ 1.50 VARIANCE OF LEFT = 1 WITH WEIGHT AND VARIANCE OF RIGHT = 0.2500 0.1111
SAMPLE SIZE = 300 WITH 75 SAMPLES ON RIGHT
DISCRETIZED MAXIMUM LIKELIHOOD PENALIZED ESTIMATE VS.
WITH WEIGHING PARAMETER ALPHA 0.1000000000D+03
41 MESH POINTS FROM -5.20000 TO 2.80000
DISCRETE MESH INTERVAL 0.20000

INTEGRATED MEAN SQUARE ERROR   0.1059614190D-01
INTEGRATED SQUARE ERROR   0.4733490400D-01
MAXIMUM ABSOLUTE DIFFERENCE   0.2316315958D+00
LOG LIKELIHOOD TERM   -0.5875622343D+03
LOG PENALTY TERM   -0.5877577822D+02

```
ABSCISSA      * ORDINATE   □ ORDINATE
              MIN 0.0      0 0.0                      MAX 0.50

-5.00D+00     6.55D-04     0.0
-4.80D+00     1.29D-03     3.46D-26
-4.60D+00     2.45D-03     7.51D-04
-4.40D+00     4.46D-03     5.30D-03
-4.20D+00     7.82D-03     1.52D-02
-4.00D+00     1.31D-02     3.07D-02
-3.80D+00     2.12D-02     5.12D-02
-3.60D+00     3.30D-02     7.64D-02
-3.40D+00     4.92D-02     1.06D-01
-3.20D+00     7.05D-02     1.41D-01
-3.00D+00     9.71D-02     1.79D-01
-2.80D+00     1.29D-01     2.18D-01
-2.60D+00     1.63D-01     2.57D-01
-2.40D+00     2.00D-01     2.93D-01
-2.20D+00     2.34D-01     3.23D-01
-2.00D+00     2.64D-01     3.45D-01
-1.80D+00     2.86D-01     3.57D-01
-1.60D+00     2.98D-01     3.60D-01
-1.40D+00     2.86D-01     3.51D-01
-1.20D+00     2.64D-01     3.32D-01
-1.00D+00     2.34D-01     3.03D-01
-8.00D-01     2.00D-01     2.65D-01
-6.00D-01     1.63D-01     2.19D-01
-4.00D-01     1.29D-01     1.69D-01
-2.00D-01     9.72D-02     1.17D-01
 0.0          7.07D-02     6.81D-02
 2.00D-01     5.05D-02     2.65D-02
 4.00D-01     5.42D-02     4.05D-03
 6.00D-01     4.08D-02     1.69D-18
 8.00D-01     1.10D-01     5.69D-03
 1.00D+00     2.91D-01     3.32D-17
 1.20D+00     2.88D-01     2.99D-02
 1.40D+00     2.01D-01     6.16D-02
 1.60D+00     9.78D-02     8.02D-02
 1.80D+00     3.33D-01     6.17D-02
 2.00D+00     7.96D-03     4.29D-02
 2.20D+00     1.78D-04     1.77D-02
 2.40D+00     1.36D-03     0.0
 2.60D+00     1.78D-04     0.0
 2.80D+00     5.45D-05     0.0
 3.00D+00     1.45D-05     0.0
 3.20D+00     1.76D-06     0.0
 3.40D+00     6.73D-07     0.0
 3.60D+00     2.39D-07     0.0
 3.80D+00     8.26D-08     0.0
 4.00D+00     2.38D-08     0.0
 4.20D+00     6.26D-09     0.0
 4.40D+00     2.49D-09     0.0
 4.60D+00     6.00D-10     0.0
 4.80D+00     2.00D-10     0.0
 5.00D+00     2.00D-10     0.0
END OF PLOT
```

Figure 5.2. $n = 300$ Bimodal DMPLE $\alpha = 10^2$.

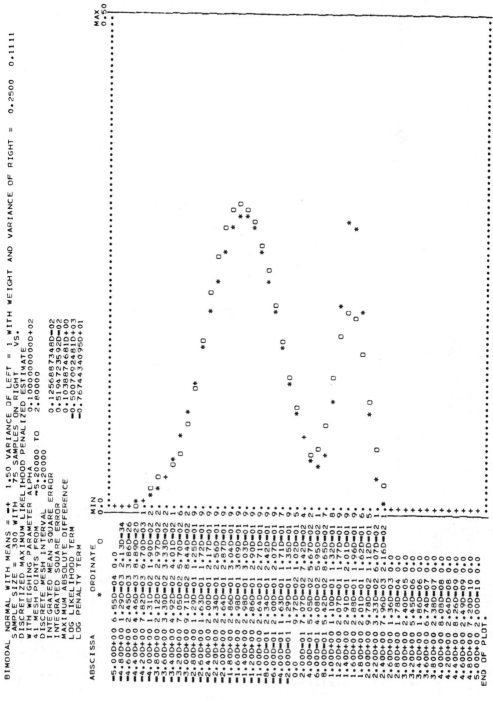

Figure 5.3. $n = 300$ Bimodal DMPLE $\alpha = 10$.

134

BIMODAL NORMAL WITH MEANS -1,+1 1.50 VARIANCE OF LEFT = 1 WITH WEIGHT AND VARIANCE OF RIGHT = 0.2500 0.1111

```
SAMPLE SIZE = 300 WITH 75 SAMPLES ON RIGHT VS.
DISCRETIZED MAXIMUM LIKELIHOOD PENALIZED ESTIMATE
WITH WEIGHING PARAMETER ALPHA     0.100000000D+01
41 MESH POINTS FROM   -5.20000 TO   2.80000
DISCRETE MESH INTERVAL   0.20000
INTEGRATED MEAN SQUARE ERROR      0.4046088973D-03
INTEGRATED SQUARE ERROR           0.2044613751D-02
MAXIMUM ABSOLUTE DIFFERENCE       0.449957615D-01
LOG LIKELIHOOD TERM               0.4940368135D+03
LOG PENALTY TERM                 -0.3057903913D+01
```

```
ABSCISSA     ORDINATE
   *            O

                            MIN              MAX
                            0.0              0.50

-5.00D+00   6.55D-04   0.0
-4.80D+00   1.29D-03   3.25D-66
-4.60D+00   2.45D-03   8.23D-50
-4.40D+00   4.46D-03   1.93D-39
-4.20D+00   7.82D-03   5.50D-04
-4.00D+00   1.31D-02   2.31D-02
-3.80D+00   2.12D-02   4.10D-02
-3.60D+00   3.30D-02   2.89D-02
-3.40D+00   4.92D-02   3.09D-02
-3.20D+00   7.05D-02   4.95D-02
-3.00D+00   9.71D-02   6.34D-02
-2.80D+00   1.29D-01   1.09D-01
-2.60D+00   1.63D-01   1.21D-01
-2.40D+00   2.34D-01   1.79D-01
-2.20D+00   2.64D-01   2.21D-01
-2.00D+00   2.86D-01   2.59D-01
-1.80D+00   2.98D-01   3.12D-01
-1.60D+00   2.86D-01   3.07D-01
-1.40D+00   2.64D-01   2.70D-01
-1.20D+00   2.34D-01   2.72D-01
-1.00D+00   2.00D-01   2.43D-01
-8.00D-01   1.63D-01   1.99D-01
-6.00D-01   1.29D-01   1.37D-01
-4.00D-01   9.72D-02   1.08D-01
-2.00D-01   7.07D-02   8.06D-02
 0.0        5.05D-02   2.59D-02
 2.00D-01   4.08D-02   4.15D-02
 4.00D-01   5.42D-02   1.25D-01
 6.00D-01   1.10D-01   2.18D-01
 8.00D-01   2.07D-01   2.65D-01
 1.00D+00   2.91D-01   2.51D-01
 1.20D+00   2.80D-01   1.81D-01
 1.40D+00   2.01D-01   1.04D-01
 1.60D+00   9.78D-02   4.35D-02
 1.80D+00   3.33D-02   1.73D-02
 2.00D+00   7.96D-03   0.0
 2.20D+00   1.36D-03   0.0
 2.40D+00   1.78D-04   0.0
 2.60D+00   5.45D-05   0.0
 2.80D+00   1.80D-06   0.0
 3.00D+00   6.74D-06   0.0
 3.20D+00   2.38D-07   0.0
 3.40D+00   8.08D-08   0.0
 3.60D+00   2.60D-08   0.0
 3.80D+00   8.25D-09   0.0
 4.00D+00   2.49D-09   0.0
 4.20D+00   7.20D-10   0.0
 5.00D+00   2.00D-10   0.0
END OF PLOT
```

Figure 5.4. $n = 300$ Bimodal DMPLE $\alpha = 1$.

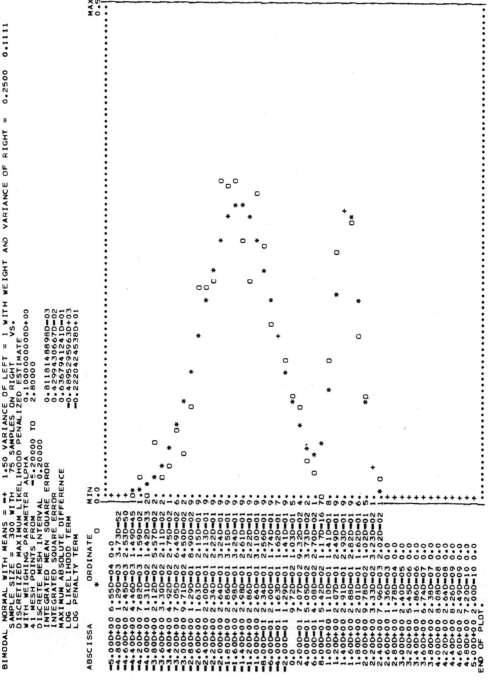

Figure 5.5. $n = 300$ Bimodal DMPLE $\alpha = 10^{-1}$.

BIMODAL NORMAL WITH MEANS = -+ 1.50 VARIANCE OF LEFT = 1 WITH WEIGHT AND VARIANCE OF RIGHT = 0.2500 0.1111
SAMPLE SIZE = 300 WITH 75 SAMPLES ON RIGHT VS.
DISCRETIZED MAXIMUM LIKELIHOOD PENALIZED ESTIMATE
WITH WEIGHING PARAMETER ALPHA 0.10000000D-01
41 MESH POINTS FROM -5.20000 TO 2.80000
DISCRETE MESH INTERVAL 0.20000
INTEGRATED MEAN SQUARE ERROR 0.1900463957D-02
INTEGRATED SQUARE ERROR 0.9000829638D-02
MAXIMUM ABSOLUTE DIFFERENCE 0.1182738981D+00
LOG LIKELIHOOD DIFFERENCE -0.4871999786D+03
LOG PENALTY TERM -0.8969259766D+00

MIN 0.0 MAX 0.50

ABSCISSA	ORDINATE *	O
-5.00D+00	6.55D-04	0.0
-4.80D+00	1.29D-03	4.81D-45
-4.60D+00	2.45D-03	1.05D-43
-4.40D+00	4.46D-03	3.26D-42
-4.20D+00	7.82D-03	1.67D-02
-4.00D+00	1.31D-02	8.82D-16
-3.80D+00	2.12D-02	6.21D-02
-3.60D+00	3.30D-02	1.87D-02
-3.40D+00	4.92D-02	2.60D-02
-3.20D+00	7.05D-02	7.26D-02
-3.00D+00	9.71D-02	4.02D-02
-2.80D+00	1.29D-01	7.42D-02
-2.60D+00	1.63D-01	2.49D-01
-2.40D+00	2.00D-01	2.16D-01
-2.20D+00	2.34D-01	1.71D-01
-2.00D+00	2.64D-01	3.76D-01
-1.80D+00	2.86D-01	3.57D-01
-1.60D+00	2.98D-01	2.64D-01
-1.40D+00	2.86D-01	1.68D-01
-1.20D+00	2.64D-01	3.67D-01
-1.00D+00	2.34D-01	2.47D-01
-8.00D-01	2.00D-01	1.52D-01
-6.00D-01	1.63D-01	1.72D-01
-4.00D-01	1.29D-01	1.44D-01
-2.00D-01	9.72D-02	9.31D-02
0.0	7.07D-02	1.03D-01
2.00D-01	5.05D-02	2.86D-02
4.00D-01	4.08D-02	3.20D-02
6.00D-01	5.42D-02	1.24D-16
8.00D-01	1.10D-01	1.49D-01
1.00D+00	2.07D-01	2.53D-01
1.20D+00	2.91D-01	2.74D-01
1.40D+00	2.88D-01	3.23D-01
1.60D+00	2.01D-01	1.20D-01
1.80D+00	9.78D-02	1.18D-01
2.00D+00	3.33D-02	2.52D-02
2.20D+00	7.96D-03	2.17D-02
2.40D+00	1.36D-03	0.0
2.60D+00	1.78D-04	0.0
2.80D+00	1.40D-04	0.0
3.00D+00	5.40D-05	0.0
3.20D+00	4.50D-06	0.0
3.40D+00	1.65D-06	0.0
3.60D+00	6.78D-07	0.0
3.80D+00	8.08D-08	0.0
4.00D+00	2.64D-08	0.0
4.20D+00	8.26D-09	0.0
4.40D+00	2.49D-09	0.0
4.60D+00	7.20D-10	0.0
4.80D+00	2.00D-10	0.0
5.00D-02		0.0

END OF PLOT.

Figure 5.6. $n = 300$ Bimodal DMPLE $\alpha = 10^{-2}$.

137

when $\alpha = 1$, we see the hint of the beginning of unstable behavior. This is still more in evidence in Figure 5.5, when $\alpha = .1$. Figure 5.6, with $\alpha = .01$, shows a distinct beginning of Dirac degeneracy. Probably, a user, reviewing the effects of the six α choices (of course, without the benefit of the asterisks of the true density) would opt for an α of 10 or 1, since one should seek the highest resolution consistent with stability.

Now, in order to compare the approximation properties of the DMPLE with those of kernel estimators and the robustness of the choice of penalty parameter α in DMPLE with that of the scaling parameter h in kernel estimation, we performed a Monte Carlo simulation. A "reasonable" value for α was chosen for each density (e.g., $\alpha = 30$ for the bimodal Gaussian). For comparison purposes we weighted the simulation study in favor of the kernel estimator by using the optimal choice of h (since in this case the solution is known). The highly popular Gaussian kernel

$$(39) \qquad K(x) = \frac{1}{\sqrt{2\pi}} \exp\left(-\frac{x^2}{2}\right)$$

was used, although kernels with finite support enjoy computational savings. The optimal choice for the scaling parameter h as a function of n in this case is

$$(40) \qquad h(n) = \left\{ \frac{\int K(x)^2 \, dx}{\left[\int x^2 K(x) \, dx\right]^2 \int [f''(x)]^2 \, dx} \right\}^{1/5} n^{-1/5}.$$

Random samples were generated on the computer and the integrated mean square error (IMSE) evaluated numerically. The Monte Carlo technique is to report the mean and standard deviation of the IMSE of 25 generated samples from a fixed distribution for a fixed sample size n. We also calculated the kernel estimate (using the "optimal" choice of h) for the same random samples and evaluated the IMSE numerically in the same manner. These results are given in Table 5.1.

We now consider the sensitivity or robustness of these estimators with respect to the parameter α for the DMPLE and the parameter $h(n)$ for the Gaussian kernel estimator. These results are given in Table 5.2.

In Table 5.2, the same normal samples generated for Table 5.1 were used with values of α and $h(n)$ perturbed by a factor of 2. Clearly, as n increases, the DMPLE estimates are insensitive to values of α in this range; however,

Table 5.1   Monte Carlo Estimation of Integrated Mean Square Error of DMPLE and Gaussian Kernel Estimator*

Sampling density	Sample size	DMPLE	Gaussian kernel
$N(0, 1)$	$n = 25$	.010 (.008)	.016 (.012)
$N(0, 1)$	$n = 100$	.0037 (.0021)	.0050 (.0027)
$N(0, 1)$	$n = 400$	.0015 (.0008)	.0020 (.0009)
bimodal	$n = 25$	.010 (.003)	.009 (.007)
bimodal	$n = 100$	.0036 (.0007)	.0036 (.0020)

* Each row represents the mean of the IMSE for 25 trials of the DMPLE and the Gaussian kernel estimator based on 25 random samples from the density in question for fixed n, the standard error of these 25 IMSE is given in parentheses, $\alpha = 10$ was used for the $N(0, 1)$, $\alpha = 30$ was used for the bimodal and the bimodal density is the mixture $.5N(-1.5, 1) + .5N(1.5, 1)$.

the integrated mean square error of the kernel estimates varies by a factor of 2. The sensitivity of the kernel estimator can also be derived from asymptotic arguments. If we rewrite equation (40) as $h(n) = Cn^{-1/5}$, then from (140) of Chapter 2 we see that

$$(41) \qquad \text{IMSE} \approx \left(\frac{\theta}{C} + \mu C^4 \right) n^{-4/5},$$

Table 5.2.   Monte Carlo Sensitivity Analysis of α and $h(n)$*

Sample size	DMPLE			Gaussian kernel		
	$\alpha = 5$	$\alpha = 10$	$\alpha = 20$	$.5h(N)$	$h(N)$	$2h(N)$
$n = 25$	.011	.010	.014	.030	.016	.031
$n = 100$	.0043	.0037	.0039	.0103	.0050	.0136
$n = 400$	.0016	.0015	.0014	.0032	.0021	.0067

* The samples used in this table are the same random samples used in Table 5.1 from the sampling density $N(0, 1)$. The table values, as in Table 5.1, represent the means of the IMSE obtained from these 25 trials.

where θ and μ are constants depending on K and f, thus perturbing $h(n)$ (and hence C) by a factor β results in a change in IMSE of β^{-1} or β^4, depending on the relative magnitude of θ and μ. In general, the design parameter of discrete maximum penalized likelihood estimation appears more insensitive to changes of sample size than the kernel width required for kernel density estimation.

Next we use Monte Carlo methods to estimate the rate of convergence of the DMPLE as a function of n, using Gaussian random samples. When plotted on log–log paper, the values in Table 5.3 fall on a straight line with slope $-.773$. The actual regression analysis gave

$$(42) \qquad \log_{10}(\text{IMSE}) = -.773 \log_{10} n - .873,$$

with a sample correlation of $r = -.996$. Thus in this case the IMSE $\approx$ $O(n^{-.773})$, which is about the same as that for kernel estimators, namely, $O(n^{-4/5})$ as was discussed in Section 2.5 of Chapter 2.

Table 5.3 Asymptotic Rate of Convergence for DMPLE Based on the $N(0, 1)$ Sampling Density

Sample size	Number of samples	Mean IMSE
25	50	.0110
100	100	.00347
400	50	.00151
800	50	.000843
1,000	50	.000545
2,000	84	.000360

In Figure 5.7, we show the DMPLE (hollow squares) for the density of annual snowfall in Buffalo, New York. Although the data was merely preprocessed by a scale and translation to lie between -3.3 and 3.3 and a "rough and ready" α of 1 was used, the fit agrees very well with that obtained by an interactive mode using older techniques.

There are a number of practical procedures for selecting α. For example, we might rescale and translate the data so that the 5 percentile lies at -3, the 95 percentile lies at $+3$, and use $\alpha = 1$. The important fact is that numerical experimentation has led us to conjecture that α selection is not as fragile a design consideration as parameter selection in state-of-the-art procedures.

```
SAMPLE SIZE =    63
DISCRETIZED MAXIMUM LIKELIHOOD PENALIZED ESTIMATE
WITH WEIGHING PARAMETER ALPHA          0.1000000000D+01
67 MESH POINTS FROM        -3.30000 TO    3.30000
DISCRETE MESH INTERVAL    0.10000
LOG LIKELIHOOD TERM                    -0.1039526466D+03
LOG PENALTY TERM                       -0.1899760704D+01
```

```
ABSCISSA  ORDINATE MIN                                                      MAX
               0    0.0                                                     0.4
                        ..................................................
-3.65D+00 0.00D+00 IO
-3.45D+00 0.00D+00 IO
-3.25D+00 0.00D+00 IO
-3.05D+00 1.48D-02 1.     O
-2.85D+00 1.16D-02 I.   O
-2.65D+00 5.31D-03 I.  O
-2.45D+00 1.07D-02 I.  O
-2.25D+00 5.02D-02 3.        O
-2.05D+00 9.48D-02 I.            O
-1.85D+00 1.24D-01 1.              O
-1.65D+00 1.58D-01 2.                  O
-1.45D+00 1.83D-01 3.                    O
-1.25D+00 1.88D-01 3.                     O
-1.05D+00 1.76D-01 2.                   O
-8.50D-01 1.68D-01 1.                  O
-6.50D-01 1.82D-01 2.                   O
-4.50D-01 2.16D-01 1.                      O
-2.50D-01 2.63D-01 5.                          O
-5.00D-02 3.05D-01 2.                             O
 1.50D-01 3.34D-01 7.                               O
 3.50D-01 3.37D-01 5.                                O
 5.50D-01 3.09D-01 3.                             O
 7.50D-01 2.58D-01 5.                         O
 9.50D-01 2.00D-01 I.                    O
 1.15D+00 1.58D-01 1.                 O
 1.35D+00 1.48D-01 1.                O
 1.55D+00 1.57D-01 3.                 O
 1.75D+00 1.67D-01 2.                  O
 1.95D+00 1.71D-01 3.                   O
 2.15D+00 1.68D-01 2.                  O
 2.35D+00 1.53D-01 1.                 O
 2.55D+00 1.29D-01 2.              O
 2.75D+00 9.80D-02 1.            O
 2.95D+00 5.59D-02 1.        O
 3.15D+00 8.60D-03 I. O
 3.35D+00 0.00D+00 IO
 3.55D+00 0.00D+00 IO
END OF PLOT.          ..................................................
```

Figure 5.7. Snowfall Densities in Buffalo, New York, over a Period of 67 Years.

Finally, we consider an extension of discrete maximum penalized likelihood to the estimation of multivariate densities, using a pseudo-independence algorithm given in [1] and [2]. Let x be a p dimensional random variable with unknown density f. Suppose we have a random sample $\{x_1, x_2, \ldots, x_n\}$ with sample mean $\bar{x}$ and positive definite sample covariance matrix $\sum$. Let Λ denote the $p \times p$ diagonal matrix, where the diagonal consists of the eigenvalues of $\sum$, and let E denote the corresponding $p \times p$ matrix whose

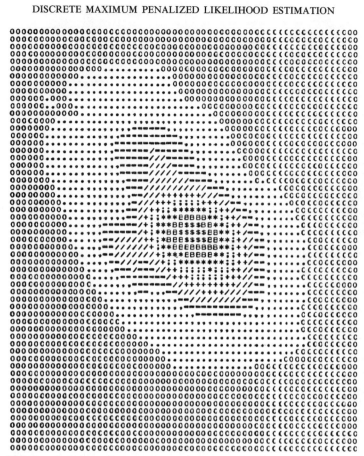

Figure 5.8. Pseudo-Independent Discrete Estimate of Reflective Intensities of Corn Data in the .40–.44 μm × .72–.80 μm Bands.

columns are the normalized eigenvectors. We take

$$(43) \qquad\qquad R = \Lambda^{-1/2} E.$$

Then,

$$(44) \qquad\qquad z = R(x - \bar{x})$$

has mean 0 and identity covariance matrix. We then use our one dimensional DMPLE algorithm on the quasi-independent components of z. The resulting density may then be transformed back into x space.

This procedure is only theoretically justified for Gaussian data. Interestingly, we have not yet found a real data set which works as a counter-example to the algorithm, although hypothetical counterexamples may be

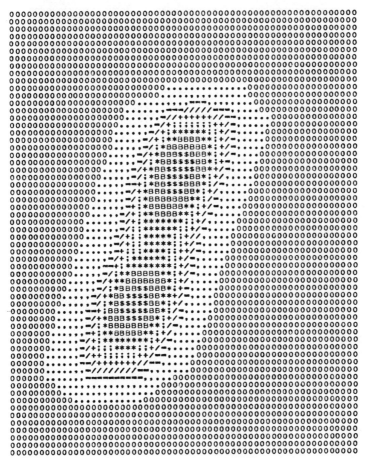

Figure 5.9. Pseudo-Independent Discrete Estimate of Reflective Intensities of Corn Data in the .66–.72 μm × .72–.80 μm Bands.

constructed quite easily. In Figure 5.8 we see the result of applying the quasi-independence algorithm to two channels of a multispectral scanner viewing the reflective light intensity from a corn field in the .40–.44 μm (abscissa) × .72–.80 μm bands. The DMPLE fit is quite satisfactory, but we might have done as well using the standard NASA LARYS algorithm which assumes Gaussianity. In Figures 5.9 through 5.11 we see the DMPLE fit for intensities in .66–.72 μm × .72–.80 μm, .66–.72 μm × .80 × 1.00 μm, and .72–.80 μm × .80–1.00 μm bands respectively. For these far more typical (of crop remote sensing data) data sets, the DMPLE procedure easily exhibits the multimodal density structure, unlike the standard parametric procedure.

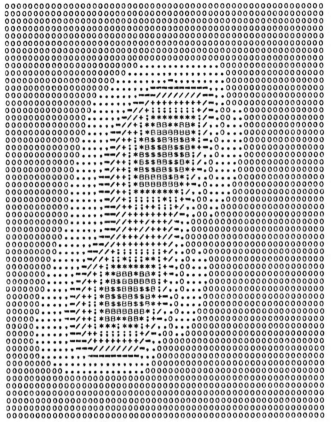

Figure 5.10. Pseudo-Independent Discrete Estimate of Reflective Intensities of Corn Data in the .66–.72 μm × .80–1.00 μm Bands.

References

[1] Bennett, J. O. (1974). "Estimation of a multivariate probability density function using B-splines." Doctoral dissertation at Rice University, Houston, Texas.

[2] Bennett, J. O., de Figueiredo, R. J. P., and Thompson, J. R. (1974). Classification by means of B-spline potential functions with applications to remote sensing. *The Proceedings of the Sixth Southwestern Symposium on System Theory*, FA3.

[3] Chung, K. L. (1968). *A Course in Probability Theory*. New York: Harcourt Brace and World.

[4] Subroutine NDMPLE, International Mathematical and Statistical Libraries, Houston, Texas.

[5] Scott, D. W. (1976). "Nonparametric probability density estimation by optimization theoretic techniques." Doctoral dissertation at Rice University, Houston, Texas.

[6] Scott, D. W., Tapia, R. A., and Thompson, J. R. (1976). "An algorithm for density estimation." *Computer Science and Statistics: Ninth Annual Symposium on the Interface*, Harvard University, Cambridge, Massachusetts.

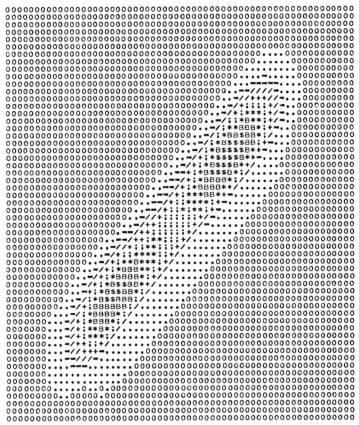

Figure 5.11. Pseudo-Independent Discrete Estimates of Reflective Intensities of Corn Data in the .72–.80 μm × .80–1.00 μm Bands.

[7] ——— (1977). "Nonparametric probability density estimation by discrete maximum penalized likelihood criteria" (submitted for publication).

[8] ——— (1977). Kernel density estimation revisited. *Nonlinear Analysis* 1: 339–72.

[9] Tapia, R. A. (1974). "A stable approach to Newton's method to general mathematical programming problems in R^n." *Journal of Optimization Theory and Application* 14: 453–76.

[10] ——— (1977). "Diagonalized multiplier methods and quasi-Newton methods for constrained optimization." *Journal of Optimization Theory and Application* 22: 135–94.

[11] Wahba, G. (1976). "A survey of some smoothing problems and the method of generalized cross-validation for solving them." Department of Statistics, University of Wisconsin–Madison TR #457. To appear, Conference on the Applications of Statistics, held at Dayton, Ohio, June 14–17, 1976, P. R. Krishnaiah, ed.

I

An Introduction
to Mathematical Optimization
Theory

I.1. Hilbert Space

In this section we review some of the basic properties of Hilbert spaces. For our purposes all vector spaces are real vector spaces. We only prove results which cannot be immediately found in standard analysis texts, for example, Dunford and Schwartz [2], Goffman and Pedrick [3], Royden [11], and Taylor [13]. For related texts with a view toward applications the reader is also referred to Daniel [1], Goldstein [4], Kantorovich and Akilov [5], Luenberger [7], and Rall [10].

Definition 1. The pair $(H, \langle \cdot, \cdot \rangle)$ is called an *inner product space* if H is a vector space and $\langle \cdot, \cdot \rangle : H \times H \to R$ satisfies the properties:
  (i) $\langle x, x \rangle \geq 0$ with equality if and only if $x = 0$.
  (ii) $\langle x, y \rangle = \langle y, x \rangle \ \forall x, y \in H$.
  (iii) $\langle \alpha x + \beta y, z \rangle = \alpha \langle x, z \rangle + \beta \langle y, z \rangle \ \forall \alpha, \beta \in R$ and $\forall x, y, z \in H$.

Definition 2. By the norm on the inner product space H we mean $\| \cdot \| : H \to R$ defined by

$$(1) \qquad \qquad \|x\| = \langle x, x \rangle^{1/2}$$

For the sake of simplicity, when the inner product is understood, the inner product space $(H, \langle \cdot, \cdot \rangle)$ is simply denoted by H. Moreover, when it is necessary to differentiate between norms or inner products in different spaces, we will write $\| \cdot \|_H$ or $\langle \cdot, \cdot \rangle_H$.

Proposition 1. (properties of inner product and norm.)
  (i) $\|x\| \geq 0$ with equality if and only if $x = 0$.
  (ii) $\|\alpha x\| = |\alpha| \|x\|, \ \forall \alpha \in R$ and $\forall x \in H$.

146

(iii) $|\langle x, y \rangle| \leq \|x\| \, \|y\|$ $\forall x, y \in H$ (Cauchy–Schwarz).

(iv) $|\, \|x\| - \|y\| \, | \leq \|x + y\| \leq \|x\| + \|y\|$ $\forall x, y \in H$ (triangle inequality).

Definition 3. When H and J are inner product spaces, then $f: H \to J$ is called an *operator* and, if $J = R$, the operator f is said to be a *functional*. Moreover, the operator f is said to be *linear* if

$$f(\alpha x + \beta y) = \alpha f(x) + \beta f(y), \qquad \forall \alpha, \beta \in R \quad \text{and} \quad \forall x, y \in H.$$

The norm on H is not a linear functional; however, the functional $f(\cdot) = \langle x, \cdot \rangle$ for fixed $x \in H$ is linear.

Let H be an inner product space.

Definition 4. A sequence $\{x^m\} \subset H$ is said to converge to $x^* \in H$ (denoted $x^m \to x$), if given $\epsilon > 0$ there exists an integer N, such that $\|x^m - x^*\| \leq \epsilon$ whenever $m \geq N$.

Definition 5. A sequence $\{x^m\} \subset H$ is said to be a *Cauchy sequence* if given $\epsilon > 0$ there exists an integer N, such that $\|x^n - x^m\| \leq \epsilon$ whenever $n, m \geq N$.

Definition 6. An inner product space H is said to be *complete* if every Cauchy sequence in H converges to a member of H. A complete inner product space is called a *Hilbert space*.

By the *dimension* of an inner product space we mean the vector space dimension, i.e., the cardinality of an algebraic (Hamel) basis. As we shall see, the existence theory for solutions of optimization problems is very much dependent on the Hilbert space (completeness) structure. The jth derivative of f is denoted by $f^{(j)}$ with $f^{(0)}$ denoting f. On occasion we use f' for $f^{(1)}$ and f'' for $f^{(2)}$.

Example 1. The vector space

$$R^q = \{x = (x_1, \ldots, x_q) : x_i \in R\}$$

with inner product

$$(2) \qquad \langle x, y \rangle = \sum_{i=1}^{q} x_i y_i$$

is a q-dimensional Hilbert space.

Example 2. The vector space

$$L^2(a, b) = \{f : (a, b) \to R : f \text{ is Lebesgue square integrable}\}$$

with inner product

(3) $$\langle f, g \rangle = \int_a^b f(t)g(t)\, dt$$

is an infinite dimensional Hilbert space if we identify members of $L^2(a, b)$, which differ on a set of Lebesgue measure zero.

Example 3. The vector space

$$C(a, b) = \{\text{all continuous } f : (a, b) \to R\}$$

with inner product given by (3) is an infinite dimensional inner product space.

Example 4 (Sobolev spaces on the real line.)   For $s = 1, 2, \ldots$ the vector space

$$H^s(-\infty, \infty) = \{f : f^{(j)} \in L^2(-\infty, \infty) \text{ for } j = 0, \ldots, s\}$$

with inner product

$$\langle f, g \rangle = \sum_{j=0}^{s} \langle f^{(j)}, g^{(j)} \rangle_{L^2(-\infty, \infty)}$$

is an infinite dimensional Hilbert space. We may think of $H^0(-\infty, \infty)$ as $L^2(-\infty, \infty)$.

Example 5 (Restricted Sobolev Spaces.)   Let (a, b) be a finite interval. Then the vector space

$$H_0^S(a, b) = \{f : f^{(j)} \in L^2(a, b), j = 0, \ldots, s \qquad \text{and}$$
$$f^{(j)}(a) = f^{(j)}(b) = 0, j = 0, \ldots, s - 1\}$$

with inner product

(5) $$\langle f, g \rangle = \langle f^{(s)}, g^{(s)} \rangle_{L^2(a, b)}$$

is an infinite dimensional Hilbert space.

Remark.   Actually, the derivatives $f^{(j)}$ in the definition of the Sobolev spaces are taken in the sense of distributions. The main consequence of this for our purposes is that the sth derivative only exists almost everywhere (i.e., on the complement of a set of Lebesgue measure zero); however, the function and its first $s - 1$ derivatives will be absolutely continuous. For further details see Lions and Magenes [6] or Oden and Reddy [9].

Lemma 1.   The inner product space given in Example 3 is not a Hilbert space.

Proof.   Let $(a, b) = (0, 1)$ and define $\{f_n\} \subset C(0, 1)$ as follows

(6)
$$f_n(t) = \begin{cases} 0 & 0 \le t \le \dfrac{1}{2} \\ n\left(t - \dfrac{1}{2}\right) & \dfrac{1}{2} \le t \le \dfrac{1}{2} + \dfrac{1}{n} \\ 1 & \dfrac{1}{2} + \dfrac{1}{n} \le t \le 1 \end{cases}$$

It is not difficult to see that if $m \ge n$, then $\|f_n - f_m\|^2 \le \dfrac{1}{n}$. Hence, $\{f_n\}$ is a Cauchy sequence in $C(0, 1)$. However, in $L^2(0, 1)$ the sequence $\{f_n\}$ converges to the discontinuous step function

(7)
$$f^*(t) = \begin{cases} 0 & 0 \le t < \dfrac{1}{2} \\ 1 & \dfrac{1}{2} \le t < 1 \end{cases}$$

Since f^* does not differ from a continuous function on a set of Lebesgue measure zero, the sequence $\{f_n\}$ defined by (6) has no limit in $C(0, 1)$. ∎

Proposition 2.   Any finite dimensional inner product space H is a Hilbert space and is congruent to R^q ($q = $ dimension of H), i.e., there is a one-one correspondence between the elements of H and R^q which preserves the linear structure and the inner product.

Since the main existence theorem we are striving toward holds only in Hilbert space, we lose little generality by working in Hilbert space from now on. Toward this end, let H and J be Hilbert spaces.

Definition 7.   An operator $f : H \to J$ is said to be *continuous at* $x \in H$ if $\{x^n\} \subset H$ and $x^n \to x$ implies $f(x^n) \to f(x)$ in J. Moreover, f is said to be *continuous in* $S \subset H$ if it is continuous at each point in S.

Definition 8.   A linear operator $f : H \to J$ is said to be *bounded* if there exists a constant M, such that

(8)
$$\|f(x)\|_J \le M\|x\|_H, \qquad \forall x \in H.$$

The vector space of all bounded operators from H into J is denoted by $[H, J]$. We denote $[H, R]$ also by H^* and call it the *dual* or *conjugate* of H. By the *operator norm of* $f \in [H, J]$ we mean

(9)
$$\|f\|_{[H, J]} = \sup\{\|f(x)\|_J : \|x\|_H = 1\}.$$

Proposition 3.   Consider the linear operator $f : H \to J$. Then the following are equivalent:

    (i) f is bounded.

    (ii) f is continuous in H.

    (iii) f is continuous at one point in H.

The Cauchy–Schwarz inequality, (iii) of Proposition 1, shows that for fixed x contained in the Hilbert space H the linear functional $x^*(\cdot) = \langle x, \cdot \rangle$ is bounded and, consequently, $x^* \in H^*$. The following very important theorem shows that actually all members of H^* arise in this fashion.

Theorem 1 (Riesz Representation Theorem.)   If $f \in H^*$, then there exists a unique $x_f \in H$, such that

$$(10) \qquad\qquad f(x) = \langle x_f, x \rangle, \qquad \forall x \in H.$$

Moreover, the vector space H^* becomes a Hilbert space with the inner product

$$(11) \qquad\qquad \langle f, g \rangle_{H^*} = \langle x_f, x_g \rangle_H \qquad \forall f, g \in H^*,$$

and the operator norm on H^* coincides with the norm induced by the inner product (11).

Proposition 4.   All linear functionals on finite dimensional inner product spaces are continuous.

Proposition 5.   There exist linear functionals on infinite dimensional inner product spaces which are not continuous.

Proof.   Consider $C(-1, 1)$ given by Example 3. Let $\delta(f) = f(0)$ for $f \in C(-1, 1)$. Clearly, δ is a linear functional. Construct

$$(12) \qquad f_n(t) = \begin{cases} 0 & -1 \leq t \leq -\dfrac{1}{n} \\[2mm] 1 + nt & -\dfrac{1}{n} \leq t \leq 0 \\[2mm] 1 - nt & 0 \leq t \leq \dfrac{1}{n} \\[2mm] 0 & \dfrac{1}{n} \leq t \leq 1 \end{cases}$$

We have that $f_n \in C(-1, 1)$ and that $f_n \to 0$ in $C(-1, 1)$. To see this observe that

$$(13) \qquad \|f_n - 0\|^2_{C(-1, 1)} = \int_{-1}^{1} f_n(t)^2 \, dt \leq \int_{-1}^{1} f_n(t) \, dt = \frac{1}{n}.$$

However,

(14) $$1 = \delta(f_n) \nrightarrow \delta(0) = 0. \quad \blacksquare$$

Remark. Although we did not demonstrate Proposition 5 for Hilbert space it does hold in this case; however, a constructive proof is not possible.

Definition 9. A sequence $\{x^n\}$ in a Hilbert space H is said to *converge weakly* to $x^* \in H$ if $f(x^n) \to f(x^*) \, \forall f \in H^*$.

Proposition 6. If $\{x^n\}$ converges to x^* in H, then it converges weakly to x^*. Moreover, if H is finite dimensional, then $\{x^n\}$ converges to x^* if and only if $\{x^n\}$ converges weakly to x^*.

Definition 10. An operator $f: H \to J$ is said to be *weakly continuous* if $\{x^n\} \subset H$ converges weakly to $x^* \in H$, then $\{f(x^n)\}$ converges weakly to $f(x^*)$ in J.

Clearly, weak continuity implies continuity, and by definition members of H^* are weakly continuous.

Definition 11. The set $S \subset H$ is *closed*, respectively, *weakly closed*, if $\{x^n\} \subset S$ converges, respectively, converges weakly, to $x^* \in H$ implies that $x^* \in S$.

It is clear that weakly closed subsets of H are closed.

Definition 12. The set $S \subset H$ is *compact*, respectively, *weakly compact*, if $\{x^n\} \subset S$ implies that $\{x^n\}$ contains a subsequence which is convergent, respectively, weakly convergent, to a member of S.

It follows that compact subsets of H are weakly compact. However, when working with infinite dimensional Hilbert spaces, compact subsets are rare and hard to come by. Consequently, any theory which requires compactness is essentially a finite dimensional theory. In particular, we have the following characterization of compactness in finite dimensions.

Proposition 7. Consider $S \subset R^q (q < \infty)$. Then S is compact if and only if S is closed and bounded (i.e., there exists M, such that $\|s\| \leq M \, \forall s \in S$).

Proposition 7 does not hold for infinite dimensional Hilbert spaces; the missing ingredient is convexity.

Definition 13. A subset S of H is said to be *convex* if $s_1, s_2 \in S$ implies $ts_1 + (1 - t)s_2 \in S \, \forall t \in [0, 1]$.

Proposition 8. Consider convex $S \subset H$. Then S is weakly closed if S is also closed. Moreover, S is weakly compact if S is also closed and bounded.

Proposition 8 leads us to anticipate the strong role of convexity in infinite dimensional optimization problems.

I.2. Reproducing Kernel Hilbert Spaces

In statistics we are usually dealing with probability density functions which are by definition nonnegative. In order to enforce this constraint, we must work with the point evaluation functionals introduced in the previous section. Moreover, if we are to utilize compactness or weak compactness the collection of functions of interest (probability density functions) must be a closed set, which in turn essentially requires that point evaluation be a continuous operation. This is the subject of the present section. We would like to isolate function spaces which are sufficiently large or rich (i.e., infinite dimensional) and still have the property that point evaluation is a continuous operation.

Definition 14. A Hilbert space H of functions defined on a set T is said to be a *proper functional Hilbert space* if for every $t \in T$ the point evaluation functional at t is continuous, i.e., there exists M_t, such that

$$(15) \qquad |f(t)| \le M_t \|f\|, \qquad \forall f \in H.$$

Definition 15. A Hilbert space $H(T)$ of functions defined on a set T is said to be a *reproducing kernel Hilbert space* (RKHS) if there exists a *reproducing kernel functional* $K(\cdot, \cdot)$ defined on $T \times T$ with the properties
  (i) $K(\cdot, t) \in H(T), \qquad \forall t \in T$

(16)

  (ii) $f(t) = \langle f, K(\cdot, t) \rangle, \forall f \in H(T)$ and $\forall t \in T$.

Proposition 9. A Hilbert space of functions defined on a set T is a proper functional Hilbert space if and only if it is a reproducing kernel Hilbert space.

Proof. The Cauchy–Schwarz inequality (Proposition 1) shows that (ii) $\Rightarrow$ (i); while the Riesz representation theorem (Theorem 1) shows that (i) $\Rightarrow$ (ii). ∎

Proposition 9 is one of the reasons for the large emphasis on RKHS in statistical applications.

Proposition 10. The restricted Sobolev space $H_0^s(a, b)$ given in Example 5 is a RKHS.

Proof. For $f \in H_0^s(a, b)$ we may write, using Cauchy–Schwarz in $L^2(a, b)$,

$$(17) \qquad |f(t)| = \left| \int_a^t f'(u)\, du \right| \le \int_a^b |f'(u)|\, du \le (b - a)\|f'\|_{L^2(a, b)}.$$

Now, by squaring (17), integrating in t over (a, b), and taking square roots we obtain

(18) $$\|f\|_{L^2(a,\,b)} \le (b - a)^{3/2}\|f'\|_{L^2(a,\,b)}.$$

Similar arguments show that for $j = 1, 2, \ldots, s - 1$

(19) $$\|f^{(j)}\|_{L^2(a,\,b)} \le (b - a)^{3/2}\|f^{(j+1)}\|_{L^2(a,\,b)}.$$

Now, combining (17) and (19) gives

$$|f(t)| \le (b - a)^{(3s-1)/2}\|f^{(s)}\|_{L^2(a,\,b)}.$$

The result now follows from (15) and Proposition 9. ∎

Proposition 11. The integration functional defined by $Q(f) = \int_a^b f(t)\, dt$ is continuous on $H_0^s(a, b)$.

Proof. A straightforward integration by parts and the Cauchy–Schwarz inequality in $L^2(a, b)$ gives for $f \in H_0^s(a, b)$

(21) $$\left| \int_a^b f(t)\, dt \right| = \left| \int_a^b \frac{t^s}{s!} f^{(s)}(t)\, dt \right| \le \left\| \frac{t^s}{s!} \right\|_{L^2(a,\,b)} \|f^{(s)}\|_{L^2(a,\,b)}.$$

The result now follows from Proposition 3. ∎

Proposition 12. The Sobolev space $H^s(-\infty, \infty)$ given in Example 4 is a RKHS.

Proof. The proof will require the introduction of the "fractional order Sobolev spaces." Toward this end for any $s \in (-\infty, \infty)$ let

(22) $$H^s(-\infty, \infty) = \{v \in L : (1 + w^2)^{s/2}\hat{v}(w) \in L^2(-\infty, \infty)\},$$

where L is the space of Schwartz's tempered distribution and $\hat{v}$ denotes the Fourier transform of v. The inner product on $H^s(-\infty, \infty)$ is given by

(23) $$\langle u, v \rangle = \langle (1 + w^2)\hat{u}, \hat{v} \rangle_{L^2(-\infty,\,\infty)}.$$

It is shown in Lions and Magenes [6] that the inner product space $H^s(-\infty, \infty)$ given by (22) and (23) is a Hilbert space, and for $s = 1, 2, \ldots$ it is isomorphic to the Sobolev space $H^s(-\infty, \infty)$ given in Example 4. Hence, we need not worry about any ambiguity in notation. Furthermore, they show that the dual of $H^s(-\infty, \infty)$ is $H^{-s}(-\infty, \infty)$. Now, $H^s(-\infty, \infty)$ will be a RKHS if the Dirac δ distribution δ_t is in the dual. Recall that δ_t is defined to be the distribution with the property that

(24) $$u(t) = \int_{-\infty}^{\infty} \delta_t(r)u(r)\, dr.$$

The integration in (24) is taken in the sense of distributions. It follows that we want

$$(25) \qquad (1 + w^2)^{-s/2} \, \hat{\delta}_t \in L^2(-\infty, \infty).$$

Since the Fourier transform of δ_t is a constant, (25) is equivalent to

$$(26) \qquad (1 + w^2)^{-s/2} \in L^2(-\infty, \infty).$$

Clearly, (26) holds if and only if $s > \frac{1}{2}$. ∎

I.3. Convex Functionals and Differential Characterizations

Throughout this section, H will denote a Hilbert space. A majority of the following material can be found in Ortega and Rheinboldt [8]. For a more detailed account of differentiation, see also Tapia [12].

Definition 16. Let S be a convex subset of H and consider $f : H \to R$. Then

(i) f is *convex* on S if

$$(27) \quad tf(x) + (1 - t)f(y) \geq f(tx + (1 - t)y), \qquad \forall t \in [0, 1], \qquad \forall x, y \in S.$$

(ii) f is *strictly convex* if strict inequality holds in (1) for $0 < t < 1$ and $x \neq y$.

(iii) f is *uniformly convex* on S if there exists $C > 0$, such that

$$(28) \quad tf(x) + (1 - t)f(y) - f(tx + (1 - t)y) \geq Ct(1 - t)\|x - y\|^2$$

$$\forall t \in [0, 1] \quad \text{and} \quad \forall x, y \in S.$$

It should be obvious that uniform convexity implies strict convexity and that strict convexity implies convexity. We will now develop reasonable criteria for deciding when a functional is uniformly convex, strictly convex or convex.

Definition 17. Consider $f : H \to R$. Given $x, \eta_1, \ldots, \eta_n \in H$ by the nth *Gâteaux variation of S at x in the directions* $\eta_1, \ldots, \eta_n$ we mean

$$(29) \qquad f^{(n)}(x)(\eta_1, \ldots, \eta_n) = \lim_{t \to 0} t^{-1}[f^{(n-1)}(x + t\eta_n)(\eta_1, \ldots, \eta_{n-1})$$

$$- f^{(n-1)}(x)(\eta_1, \ldots, \eta_{n-1})],$$

with $f^{(0)}(x) = f(x)$. Moreover, we say f is m *times Gâteaux differentiable* in $S \subset H$ if the first m Gâteaux variations exist at x and are linear in η_i $\forall i$ and for each fixed x in S. We also say that f is *continuously differentiable* in S if it is Gâteaux differentiable in S, $f^{(1)}(x) \in H^*$ $\forall x \in S$ and $f^{(1)} : S \subset H \to H^*$

is a continuous operator. In the case that $f^{(1)}(x) \in H^*$ the Riesz representer of $f^{(1)}(x)$ is denoted by $\nabla f(x)$ and is called the *gradient* of f at x (i.e., $f^{(1)}(x)(\eta) = \langle \nabla f(x), \eta \rangle \forall \eta \in H$).

Proposition 13. If $\Phi(t) = f(x + t\eta)$, then

$$(30) \qquad \Phi^{(n)}(0) = f^{(n)}(x)(\eta, \ldots, \eta).$$

Proof. Let $n = 1$. Then

$$(31) \qquad \Phi'(t) = \lim_{\Delta t \to 0} \left[\frac{\Phi(t + \Delta t) - \Phi(t)}{\Delta t} \right]$$

$$(32) \qquad = \lim_{\Delta t \to 0} \left[\frac{f(x + t\eta + \Delta t\eta) - f(x + t\eta)}{\Delta t} \right]$$

$$(33) \qquad = f'(x + t\eta)(\eta);$$

and (30) holds for $n = 1$. Assume that (30) holds for $n = 1, \ldots, k - 1$. Then

$$(34) \quad \Phi^{(k)}(t) = \lim_{\Delta t \to 0} \left[\frac{\Phi^{(k-1)}(t + \Delta t) - \Phi^{(k-1)}(t)}{\Delta t} \right]$$

$$(35) \qquad = \lim_{\Delta t \to 0} \left[\frac{f^{(k-1)}(x + t\eta + \Delta t\eta)(\eta, \ldots, \eta) - f^{(k-1)}(x + t\eta)(\eta, \ldots, \eta)}{\Delta t} \right]$$

$$(36) \qquad = f^{(k)}(x + t\eta)(\eta, \ldots, \eta);$$

hence (30) holds for k and by induction for all n. ∎

Example 6. Consider $f : H \to R$ defined by

$$(37) \qquad f(x) = \tfrac{1}{2}\langle x, x \rangle,$$

then

$$(38) \qquad \Phi'(t) = \langle x + t\eta, \eta \rangle$$

and

$$(39) \qquad \Phi''(t) = \langle \eta, \eta \rangle.$$

It follows that $f'(x)(\eta) = \langle x, \eta \rangle$ and $f''(x)(\eta, \eta) = \langle \eta, \eta \rangle$. Observe that in this case f is infinitely Gâteaux differentiable in H.

Proposition 14. Assume that $f : H \to R$ is Gâteaux differentiable in a convex subset S of H. Then

(40) (i) f is convex in $S \Leftrightarrow f'(x)(y - x) \leq f(y) - f(x), \qquad \forall x, y \in S.$

(41) (ii) f is strictly convex in $S \Leftrightarrow f'(x)(y - x) < f(y) - f(x),$

$$\forall x, y \in S \quad \text{and} \quad x \neq y.$$

Proof [(i) ⇐].   For $x, y \in S$ let $z = \alpha x + (1 - \alpha)y$ for $0 < \alpha < 1$. By hypothesis $z \in S$ and

(42) $$f(x) - f(z) \geq f'(z)(x - z)$$

and

(43) $$f(y) - f(z) \geq f'(z)(y - z).$$

Therefore,

$$
\begin{aligned}
(44)\quad \alpha f(x) + (1 - \alpha)f(y) - f(z) &\geq \alpha f'(z)(x - z) + (1 - \alpha)f'(z)(y - z) \\
&= f'(z)[\alpha(x - z) + (1 - \alpha)(y - z)] \\
&= f'(z)[\alpha x + (1 - \alpha)y - z] \\
&= f'(z)(0) \\
&= 0.
\end{aligned}
$$

It follows that

(45) $$\alpha f(x) + (1 - \alpha)f(y) \geq f(\alpha x + (1 - \alpha)y)$$

and f is convex. Exactly the same proof demonstrates [(ii) ⇐].
[(i) ⇒].   Suppose f is convex on S. For $x, y \in S$ and $0 < \alpha < 1$ we have

(46) $$\alpha f(y) + (1 - \alpha)f(x) \geq f(\alpha y + (1 - \alpha)x).$$

Divide (46) by α and then let $\alpha \to 0$ to obtain

(47) $$f(y) - f(x) \geq f'(x)(y - x).$$

[(ii) ⇒].   Here the above limiting process will not suffice. Let $z = \frac{1}{2}(x + y)$, then $z - x = \frac{1}{2}(y - x)$ and by [(i) ⇒]

$$
\begin{aligned}
(48)\quad f'(x)(y - x) &= 2f'(x)(z - x) \\
&\leq 2[f(z) - f(x)] \\
&< 2[\tfrac{1}{2}f(x) + \tfrac{1}{2}f(y) - f(x)] \\
&= f(y) - f(x).
\end{aligned}
$$

Proposition 15.   Under the assumptions of Proposition 14, we have
  (i) f is convex on $S \Leftrightarrow [f'(y) - f'(x)](y - x) \geq 0 \qquad \forall x, y \in S$
  (ii) f is strictly convex on $S \Leftrightarrow [f'(y) - f'(x)](y - x) > 0$

$$\forall x, y \in S \quad \text{and} \quad x \neq y.$$

Proof [(i) ⇒].   Suppose f is convex on S. Then by [(i)   ] of Proposition 14

(49) $$f(y) - f(x) \geq f'(x)(y - x)$$

and

(50) $$f(x) - f(y) \geq f'(y)(x - y).$$

Now, by adding (49) to (50) we obtain

(51) $$[f'(y) - f'(x)](y - x) \geq 0.$$

$[(i) \Leftarrow]$. By the mean value theorem applied to $\Phi(t) = f(x + t(y - x))$ we have $\Phi(1) - \Phi(0) = \Phi'(t)$ for some $0 < t < 1$, or equivalently

(52) $$f(y) - f(x) = f'(x + t(y - x))(y - x), \qquad 0 < t < 1.$$

Let $u = x + t(y - x)$. Observe that $u \in S$ and that $y - x = (u - x)/t$, so that

(53) $$[f'(u) - f'(x)](y - x) = \frac{1}{t}[f'(u) - f'(x)](u - x) \geq 0.$$

Also,

(54) $$f(y) - f(x) = [f'(u) - f'(x)](y - x) + f'(x)(y - x),$$

therefore,

(55) $$f(y) - f(x) \geq f'(x)(y - x).$$

By $[(i) \Leftarrow]$ of Proposition 14 we see that (55) implies convexity. The above calculations also establish part (ii). ∎

Definition 18. Consider $f: H \to R$ and a convex subset S of H. By *the cone tangent to S at x* we mean $T(x) = \{\eta \in H : \exists t > 0, \text{ such that } x + t\eta \in S\}$. Suppose f is twice Gâteaux differentiable in S. Then f'' is said to be *positive semidefinite relative to S* if for each $x \in S$ we have

(56) $$f''(x)(\eta, \eta) \geq 0, \qquad \forall \eta \in T(x).$$

We say f'' is *positive definite relative to S* if strict inequality holds in (56) for $\eta \neq 0$. Finally, f'' is said to be *uniformly positive definite relative to S* if there exists $C > 0$, such that for each $x \in S$ we have

(57) $$f''(x)(\eta, \eta) \geq C\|\eta\|^2, \qquad \forall \eta \in T(x).$$

Proposition 16. Assume that $f: H \to R$ is twice Gâteaux differentiable in a convex subset S of H.
Then
  (i) f is convex in $S \Leftrightarrow f''$ is positive semidefinite relative to S.
  (ii) f is strictly convex in $S \Leftarrow f''$ is positive definite relative to S.

Proof [(i) ⟸]. Using Taylor's theorem on $\Phi(t) = f(x + t(y - x))$, we have for $x, y \in S$

(58) $f(y) - f(x) - f'(x)(y - x) = \frac{1}{2}f''(x + \Theta(y - x))(y - x, y - x),$

$$0 < \Theta < 1$$

Since $x + \Theta(y - x) \in S$, it follows that $(y - x) \in T(x + \Theta(y - x))$. Consequently, Proposition 14, (58) and the fact that f'' is positive semidefinite relative to S imply f is convex.

[(ii) ⟸]. This follows exactly as in the proof of [(i) ⟸].

[(i) ⟹]. First observe that if $x \in S$ and $x + t\eta \in S$, then $x + \tau\eta \in S$ for $0 \leq \tau \leq t$. For $x \in S$ and $\eta \in T(x)$ we have (considering only small $t > 0$ if necessary)

(59) $$f''(x)(\eta, \eta) = \lim_{t \to 0} \frac{1}{t} [f'(x + t\eta)(\eta) - f'(x)(\eta)]$$

$$= \lim_{t \to 0} \frac{1}{t^2} [f'(x + t\eta) - f'(x)](x + t\eta - x).$$

By convexity and [(i) ⟹] of Proposition 15 it follows that (59) implies that f'' is positive semidefinite relative to S. ∎

Proposition 17. Assume that $f : H \to R$ is Gâteaux differentiable in a convex subset S of H. Then the following three statements are equivalent:

  (i) f is uniformly convex in S with constant C,
  (ii) $f(y) - f(x) \geq f'(x)(y - x) + C\|y - x\|^2, \forall x, y \in S,$
  (iii) $(f'(y) - f'(x))(y - x) \geq 2C\|y - x\|^2, \forall x, y \in S.$

If in addition, f is twice Gâteaux differentiable in S, then the above three statements are equivalent to

  (iv) f'' is uniformly positive definite relative to S with constant $2C$.

Proof [(i)] ⟹ (ii)]. We have for $0 < t < 1$ and $x, y \in S$ from (28)

(60) $tf(y) - tf(x) \geq [f(x + t(y - x)) - f(x)] + t(1 - t)C\|x - y\|^2.$

Therefore, dividing (60) by t and then letting $t \to 0$ gives (ii).

[(ii) ⟹ (iii)]. Interchanging the roles of x and y in (ii) and adding the resulting inequality to (ii) gives (iii).

[(iii) ⟹ (ii)]. Let $t_k = \dfrac{k}{m + 1}$, $k = 0, \ldots, m + 1$ for arbitrary $m \geq 0$. The mean value theorem on $\Phi(t) = f(x + t(y - x))$ gives

(61) $\Phi(t_{k+1}) - \Phi(t_k) = \Phi'(s_k),$ where $t_k < s_k < t_{k+1}.$

Rewriting (61) leads to

(62)
$$f(x + t_{k+1}(y - x)) - f(x + t_k(y - k))$$
$$= f'(x + s_k(y - x))((t_{k+1} - t_k)(y - x)).$$

Hence, from (iii) and (62) we obtain

(63) $f(y) - f(x) = \sum_{k=0}^{m} [f(x + t_{k+1}(y - x)) - f(x + t_k(y - x))]$

$$= \sum_{k=0}^{m} [f'(x + s_k(y - x)) - f'(x)][(t_{k+1} - t_k)(y - x)]$$
$$+ f'(x)(y - x)$$

$$= \sum_{k=0}^{m} \frac{(t_{k+1} - t_k)}{s_k} [f'(x + s_k(y - x)) - f'(x)](s_k(y - x))$$
$$+ f'(x)(y - x)$$

$$\geq \sum_{k=0}^{m} \frac{2(t_{k+1} - t_k)}{s_k} s_k^2 C\|y - x\|^2 + f'(x)(y - x).$$

Now, observe that

(64)
$$\sum_{k=0}^{m} (t_{k+1} - t_k)s_k \geq \sum_{k=0}^{m} (t_{k+1} - t_k)t_k = \frac{1}{(m+1)^2} \sum_{k=0}^{m} k$$
$$= \frac{1}{2} \frac{m}{m+1}.$$

Coupling (63) with (64) we arrive at

(65)
$$f(y) - f(x) \geq \frac{m}{m+1} C\|y - x\|^2 + f'(x)(y - x).$$

Letting $m \to \infty$ in (65) gives (ii).

[(ii) $\Rightarrow$ (i)].   Let $z = \alpha x + (1 - \alpha)y$ for $x, y \in S$ and $0 < \alpha < 1$. From (ii) we have

(66)
$$f(x) - f(z) \geq f'(z)(x - z) + C\|x - z\|^2$$

and

(67)
$$f(y) - f(z) \geq f'(z)(y - z) + C\|y - z\|^2.$$

Observe that

(68)
$$\alpha\|x - z\|^2 + (1 - \alpha)\|y - z\|^2 = \alpha(1 - \alpha)\|x - y\|^2.$$

Now, multiplying (66) by α, (67) by $(1 - \alpha)$ and then adding and recalling (68) we obtain

(69) $$\alpha f(x) + (1 - \alpha)f(y) - f(z) \geq C\alpha(1 - \alpha)\|y - x\|^2.$$

However, (69) is merely (28) with t replaced by α.

$[(iv) \Rightarrow (ii)]$. Let $\Phi(t) = f(x + t(y - x))$. Taylor's theorem on $\Phi(t)$ gives

(70) $$\Phi(1) - \Phi(0) - \Phi'(0) = \tfrac{1}{2}\Phi''(\theta), \qquad \text{for} \quad 0 < \theta < 1.$$

Coupling (70) with (iv) leads to

(71) $$f(y) - f(x) - f'(x)(y - x) = \tfrac{1}{2}f''(x + \theta(y - x))(y - x, y - x)$$
$$\geq C\|x - y\|^2, \qquad \forall x, y \in S$$

which is exactly (ii).

$[(iii) \Rightarrow (iv)]$. For $x \in S$ and $\eta \in T(x)$ let $y = x + t\eta$ in (iii) to obtain

(72) $$[f'(x + t\eta) - f'(x)](t\eta) \geq 2Ct^2\|\eta\|^2.$$

Dividing (72) by t^2 and letting $t \to 0$ gives (iv). ∎

I.4. Existence and Uniqueness of Solutions for Optimization Problems in Hilbert Space

A portion of the following material can be found in Goldstein [4]. As before, H will denote a Hilbert space. For $S \subset H$ and $f : H \to R$ we may consider the constrained optimization problem

(73) $$\text{minimize } f(x); \quad \text{subject to } x \in S.$$

Solutions of problem (73) are referred to as *minimizers*.

Theorem 2 (uniqueness). If f is strictly convex and S is convex, then problem (73) has at most one minimizer.

Proof. Suppose x^* and x^{**} are both solutions of problem (73). Then, if $x^* \neq x^{**}$ we have for $0 < t < 1$ by strict convexity

(74) $$f(tx^* + (1 - t)x^{**}) < tf(x^*) + (1 - t)f(x^{**}) = f(x^*).$$

But $x = tx^* + (1 - t)x^{**} \in S$, since S is convex. This contradicts the optimality of x^*. ∎

The problem of existence of solutions to problem (73) is not as straightforward and requires the introduction of the notion of weak lower semicontinuity. Toward this end recall that the functional f is (weakly) continuous at x if given a sequence $\{x^n\} \subset H$ which converges (weakly) to $x \in H$, then given $\epsilon > 0$ there exists an integer N_ϵ, such that

$$(75) \qquad -\epsilon < f(x^n) - f(x) < \epsilon, \qquad \forall n \geq N_\epsilon.$$

Definition 19. The functional f is said to be (*weakly*) *lower semicontinuous* at $x \in H$ if given a sequence $\{x^n\} \subset H$ converging (weakly) to x, then, given $\epsilon > 0$, there exists an integer N_ϵ, such that

$$(76) \qquad -\epsilon < f(x^n) - f(x), \qquad \forall n \geq N_\epsilon.$$

Moreover, we say f is (*weakly*) *upper semicontinuous* at $x \in H$ if in the above we replace (76) with

$$(77) \qquad f(x^n) - f(x) < \epsilon, \qquad \forall n \geq N_\epsilon.$$

Clearly, f is (weakly) continuous if and only if it is both (weakly) lower and upper semicontinuous.

Proposition 18. The functional f is (weakly) lower semicontinuous if and only if whenever $\{x^n\}$ converges (weakly) to x we have

$$(78) \qquad f(x) \leq \varliminf_n f(x^n).$$

Proof. The proof is a straightforward application of (76) and the definition

$$(79) \qquad \varliminf_n f(x^n) = \lim_{n \to \infty} \left(\inf_{k \geq n} f(x^k) \right). \qquad \blacksquare$$

Proposition 19. Let D be a subset of H. Then $f : D \to R$ is (weakly) lower semicontinuous in D if and only if the set

$$D_M = \{x \in D : f(x) \leq M\}$$

is (weakly) closed for all M.

Proof. Suppose that D_M is (weakly) closed $\forall M$. Consider $x \in D$ and assume that f is not (weakly) lower semicontinuous at x, then for some M and some sequence $\{x^n\}$ converging (weakly) to x we have

$$(80) \qquad \varliminf_n f(x^n) < M < f(x).$$

It follows that a subsequence of $\{x^n\}$ remains in D_M, and since D_M is (weakly) closed, we have $x \in D_M$. However, this contradicts (80).

Now assume that f is (weakly) lower semicontinuous and that for some M the set D_M is not (weakly) closed. Then there exists $\{x^n\} \subset D_M$, such that $\{x^n\}$ converges (weakly) to $x \notin D_M$.
This implies that $f(x) > M$. However, by Proposition 18 we have that $f(x) \leq \varliminf_n f(x^n) \leq M$. This is a contradiction. $\blacksquare$

Proposition 20. A continuous convex functional f defined on a closed convex subset S of H is weakly lower semicontinuous.

Proof. The set $S_M = \{x \in S : f(x) \leq M\}$ is closed, since f is continuous and S is closed. It is also convex, since S is convex and f is convex. Therefore, by Proposition 8 the set S_M is weakly closed. This means, according to Proposition 19, that f is weakly lower semicontinuous. ∎

Theorem 3 (Existence for finite dimensions.) Suppose that in problem (73) the Hilbert space H is finite dimensional, the functional f is continuous on S, and the subset S is closed and bounded. Then problem (73) has at least one minimizer.

Proof. Choose $\{x^n\} \subset S$, such that $f(x^n)$ converges to $\inf\{f(x) : x \in S\}$. By compactness of S there exists $x^* \in S$ and a subsequence $\{x^{n_i}\}$ which converges to x^*. By the continuity of f we have that $\{f(x^{n_i})\}$ converges to $f(x^*)$; hence, x^* is a minimizer to problem (73). ∎

Theorem 4 (Existence for infinite dimensions.) Suppose that in problem (73) the functional f is convex and continuous on S, and the subset S is closed, bounded, and convex. Then problem (73) has at least one minimizer.

Proof. The proof is the same as the proof of Theorem 3 if we replace continuity with the weak lower semicontinuity guaranteed by Proposition 20, compactness with the weak compactness guaranteed by Proposition 8, and then use Proposition 18. ∎

Definition 20. Consider $S \subset H$ and $f : S \to R$. We say that f has the *infinity property* in S if $\{x^n\} \subset S$ and $\|x^n\| \to \infty$ implies $f(x^n) \to +\infty$.

Theorem 5. Let S be a closed convex subset of H. If $f : S \to R$ is convex in S, continuous in S, and has the infinity property in S, then problem (73) has at least one solution.

Proof. Consider $x^0 \in S$. Let $\hat{S} = \{x \in S : f(x) \leq f(x^0)\}$. By the infinity property the set S is bounded. Moreover, by continuity $\hat{S}$ is closed. Finally, by convexity of f the set $\hat{S}$ is convex. The theorem now follows from Theorem 4, since a minimizer in $\hat{S}$ is a minimizer in S. ∎

Theorem 6. Let S be a closed and convex subset of H. If $f : S \to R$ is continuous and uniformly convex on S, then problem (73) has a unique solution.

Proof. Uniqueness follows from Theorem 2. We will show that f has the infinity property and the result will then follow from Theorem 5. There is no loss of generality in assuming $0 \in S$. For if not we could work with $S - y_0$ for some fixed $y_0 \in S$. By uniform convexity there exists $C > 0$, such that for $0 \leq t \leq 1$ and $x, y \in S$ we have

(81) $tf(x) + (1 - t)f(y) - f(tx + (1 - t)y) \geq Ct(1 - t)\|x - y\|^2.$

Choose $y = 0$ in (81). Then

(82) $tf(x) + (1 - t)f(0) \geq f(tx) + Ct(1 - t)\|x\|^2.$

From (82) it follows that

(83) $\max(f(x), f(0)) \geq f(tx) + Ct(1 - t)\|x\|^2$

Let $\hat{S} = \{x \in S : \|x\| \leq 1\}$. Observe that $\hat{S}$ is closed, convex, and bounded; hence by Theorem 4 there exists α such that $f(x) \geq \alpha \ \forall x \in \hat{S}$. Given $x \in S$, such that $\|x\| \geq 1$ choose t_x, such that $\frac{1}{2} \leq t_x\|x\| \leq 1$. Then $0 < t_x \leq 1$, and since $0 \in S$ we have $t_x x \in S$. From (83) with this x and t_x we obtain

(84) $\max(f(x), f(0)) \geq \alpha + Ct_x(1 - t_x)\|x\|^2$

$$\geq \alpha + \frac{C}{2}(\|x\| - 1).$$

Observe that (84) implies that as $\|x\| \to +\infty$ we must have $f(x) \to \infty$. ∎

The following theorem is extremely useful in applications.

Theorem 7. Let S be a closed convex subset of H. Suppose $f : H \to R$ is continuous in S, is twice Gâteaux differentiable in S, and the second Gâteaux variation is uniformly positive definite in S. Then problem (73) has a unique solution.

Proof.   The theorem follows from Theorem 6 and Proposition 17. ∎

I.5. Lagrange Multiplier Necessity Conditions

Let us consider the special case of problem (73) in I.4, where

(85) $S = \{x \in H : g_i(x) = 0, i = 1, \ldots, m\}$

and g_i are functionals on H. Problem (73) can then be written as

(86)   minimize $f(x)$;   subject to $g_i(x) = 0$, $i = 1, \ldots, m.$

We assume that $f, g_i \forall i$ are continuously differentiable in H.

Definition 21. We say that $x \in H$ is a *regular point of the constraints* g_i, $i = 1, \ldots, m$ if

$$(87) \qquad \{\nabla g_1(x), \ldots, \nabla g_m(x)\}$$

is a linearly independent set.

Theorem 8. Suppose that x^* solves problem (86) and also that x^* is a regular point of the constraints. Then there exist unique Lagrange multipliers $\lambda_1, \ldots, \lambda_m$, such that

$$(88) \qquad \nabla f(x^*) + \lambda_1 \nabla g_1(x^*) + \cdots + \lambda_m \nabla g_m(x^*) = 0.$$

Proof. For the proof of this theorem we follow Luenberger [7]. Consider the operator $G: H \to R^{m+1}$ defined by

$$(89) \qquad G(x) = (f(x), g_1(x), \ldots, g_m(x)).$$

Assume that there exists $Z \in H$ such that $\langle Z, \nabla f(x^*) \rangle \neq 0$ and $\langle Z, \nabla g_i(x^*) \rangle = 0$, $i = 1, \ldots, m$. If we write for $\eta \in H$

$$(90) \qquad JG(x^*)(\eta) = [\langle \nabla f(x^*), \eta \rangle, \langle \nabla g_1(x^*), \eta \rangle, \ldots, \langle \nabla g_m(x^*), \eta \rangle],$$

then the operator $JG(x^*)$ takes H onto R^{m+1}, since $\{\nabla f(x^*), \nabla g_1(x^*), \ldots, \nabla g_m(x^*)\}$ is also a linearly independent set. By the inverse function theorem (see Luenberger [7]) for $y \in R^{m+1}$ near $G(x^*)$ there exists $x \in H$, such that $G(x) = y$. Choosing $y = (f(x^*) - \epsilon, 0, \ldots, 0)$ for any $\epsilon > 0$, we contradict the optimality of x^*. It follows that whenever $\langle Z, \nabla g_i(x^*) \rangle = 0$, $i = 1, \ldots, m$ we necessarily have that $\langle Z, \nabla f(x^*) \rangle$ is also zero. However, this implies by a well-known result (see, for example, Theorem 3.5–C of Taylor [13]) that $\nabla f(x^*)$ must be a linear combination of $\nabla g_1(x^*), \ldots, \nabla g_m(x^*)$. The coefficients in the linear combination are the Lagrange multipliers. By regularity these multipliers are unique. ■

References

[1] Daniel, J. W. (1971). *The Approximate Minimization of Functionals*. Englewood Cliffs, New Jersey: Prentice–Hall.

[2] Dunford, H., and Schwartz, J. T. (1958). *Linear Operators, Part I*. New York: Interscience.

[3] Goffman, C., and Pedrick, G. (1965). *First Course in Functional Analysis*. Englewood Cliffs, New Jersey: Prentice–Hall.

[4] Goldstein, A. A. (1967). *Constructive Real Analysis*. New York: Harper and Row.

[5] Kantorovich, L. V., and Akilov, G. P. (1964). *Functional Analysis in Normed Spaces*. New York: Macmillan.

[6] Lions, J. L., and Magenes, E. (1972). *Non-homogeneous Boundary Value Problems and Applications*. Translated from the French by P. Kenneth. New York: Springer–Verlag.

[7] Luenberger, D. G. (1968). *Optimization by Vector Space Methods*. New York: Wiley.

[8] Ortega, J. M., and Rheinboldt, W. C. (1970). *Iterative Solution of Nonlinear Equations in Several Variables*. New York: Academic Press.

[9] Oden, J. T., and Reddy, J. N. (1976). *An Introduction to the Mathematical Theory of Finite Elements*. New York: Wiley.

[10] Rall, L. B. (1969). *Computational Solution of Nonlinear Operator Equations*. New York: Wiley.

[11] Royden, H. (1963). *Real Analysis*. New York: Macmillan.

[12] Tapia, R. A., "The differentiation and integration of nonlinear operators," in *Nonlinear Functional Analysis and Applications* (1971). Ed. L. B. Rall. New York: Academic Press.

[13] Taylor, A. E. (1958). *Introduction to Functional Analysis*. New York: Wiley.

II

Numerical Solution of Constrained Optimization Problems

II. 1. *The Diagonalized Multiplier Method*

In this section we will present a class of effective numerical algorithms for approximating solutions of constrained optimization problems called diagonalized quasi-Newton multiplier methods. These algorithms were suggested by Tapia in [2] and the reader desiring more detail is referred to that paper. The optimization problems we are concerned with in statistics require the variables to be nonnegative. In the following section of this appendix, we will describe a method for applying algorithms for problems with equality constraints to problems which also require the variables to be nonnegative. Consequently, we present the diagonalized quasi-Newton multiplier method for problems with equality constraints in this section. The reader is referred to Section II.2 for nonnegativity constraints and to [2] for the general inequality constraint.

We are interested in problem (I.86) where $H = R^n$. Specifically, consider the constrained optimization problem

(1)   minimize $f(x)$;   subject to $g_i(x) = 0$, $i = 1, \ldots, m$

where $f, g_i : R^n \to R$ have continuous second-order partial derivatives. In this case it is a simple matter to show that the gradient of f at x (see Definition I.17) is actually the vector of partial derivatives (some texts take this as the definition), i.e.,

(2) $$\nabla f(x) = \left(\frac{\partial f(x)}{\partial x_1}, \ldots, \frac{\partial f(x)}{\partial x_n} \right)^T.$$

We use the notation $\nabla^2 f(x)$ to denote the Hessian matrix of f at x, i.e.,

$$(3) \qquad \nabla^2 f(x) = \left(\frac{\partial^2 f(x)}{\partial x_i \, \partial x_j}\right) \qquad i, j = 1, \ldots, n.$$

In (3) the notation (a_{ij}) is used to denote the matrix whose (i, j)-th element is given by a_{ij}, while in (2) the superscript T, as usual, denotes transposition. For the sake of simplicity, define $g : R^n \to R^m$ by

$$(4) \qquad g(x) = (g_1(x), \ldots, g_m(x))^T.$$

It makes sense to denote the transpose of the Jacobian of g by $\nabla g(x)$, since the ith column of this matrix is merely the vector $\nabla g_i(x)$, i.e., the $n \times m$ matrix $\nabla g(x)$ can be written

$$\nabla g(x) = [\nabla g_1(x), \ldots, \nabla g_m(x)].$$

Consider the *augmented Lagrangian* for problem (1), i.e., $L : R^{n+m+1} \to R$, where

$$(5) \qquad L(x, \lambda, C) = f(x) + \lambda^T g(x) + \frac{C}{2} g(x)^T g(x)$$

for $x \in R^n$, $\lambda \in R^m$, $C \geq 0$ and g given by (4). Since the augmented Lagrangian is a function of the three variables x, λ and C, we use the notation $\nabla_x L(x, \lambda, C)$ and $\nabla_x^2 L(x, \lambda, C)$ to denote differentiation with respect to x (λ and C are held fixed). In (5) λ can be thought of as an approximate Lagrange multiplier and C as a penalty constant.

Consider a quasi-Newton method for the unconstrained optimization problem

$$(6) \qquad \underset{x}{\text{minimize }} L(x, \lambda, C)$$

with inverse Hessian update formula

$$(7) \qquad \bar{H} = \mathscr{H}(x, \lambda, C, H),$$

i.e., the iteration

$$(8) \qquad x^{k+1} = x^k - H^k \nabla_x L(x^k, \lambda, C)$$

and
$$k = 0, 1, 2, \ldots$$

$$H^{k+1} = \mathscr{H}(x^k, \lambda, C, H^k),$$

where H^k is a conscious attempt to approximate, $\nabla_x^2 L(x^k, \lambda, C)^{-1}$, the inverse of the Hessian of the augmented Lagrangian. For more information on quasi-Newton methods for unconstrained optimization the reader is referred to Dennis and Moré [1].

The two main ingredients of the diagonalized quasi-Newton multiplier method are the unconstrained quasi-Newton method described in equations (6)–(8) and the *Lagrange Multiplier update formula.*

(9) $\quad U(x, \lambda, C, H) = \lambda + (\nabla g(x)^T H \, \nabla g(x))^{-1}[g(x) - \nabla g(x)^T H \, \nabla_x L(x, \lambda, C)].$

By a *diagonalized quasi-Newton multiplier method* for problem (1) we mean

Step 1:   Determine x, λ, C and H

Step 2:

(10) $\qquad\qquad\qquad \bar{\lambda} = U(x, \lambda, C, H)$

(11) $\qquad\qquad\qquad \bar{x} = x - H \, \nabla_x L(x, \bar{\lambda}, C)$

Step 3:

(12) $\qquad\qquad\qquad \bar{H} = \mathscr{H}(x, \bar{\lambda}, C, H)$

Replace (x, λ, H) with $(\bar{x}, \bar{\lambda}, \bar{H})$ and go to Step 2.

Remark.   Clearly, in implementing this algorithm a stopping criterion of the form

$$|\bar{x}_i - x_i| \le \epsilon, \qquad i = 1, \ldots, n$$

and

$$|\bar{\lambda}_i - \lambda_i| \le \epsilon, \qquad i = 1, \ldots, m$$

for some predetermined small number ϵ would have to be inserted between Step 2 and Step 3.

It is usual to choose C to be either zero or some small number, e.g., $C = .1$. Rules for determining optimal values of C are currently under investigation.

By the *diagonalized Newton multiplier method* we mean the choice (12) is given by

(13) $\qquad\qquad\qquad \bar{H} = \nabla_x^2 L(\bar{x}, \bar{\lambda}, \bar{C})^{-1},$

which corresponds to using Newton's method as the unconstrained minimization algorithm for problem (6).

Two standard choices for *diagonalized secant multiplier methods* arise by letting $\bar{H}$ in (12) be given by the following popular secant update formulas (see Dennis and Moré [1]): *Davidon–Fletcher–Powell* (DFP)

(14) $\qquad\qquad \bar{H} = H + \dfrac{ss^T}{\langle s, y \rangle} - \dfrac{Hyy^T H}{\langle y, Hy \rangle};$

Broyden–Fletcher–Goldfarb–Shanno (BFGS)

(15) $\qquad \bar{H} = H + \dfrac{(s - Hy)s^T + s(s - Hy)^T}{\langle s, y \rangle} - \dfrac{\langle s - Hy, y \rangle ss^T}{\langle s, y \rangle^2},$

with

$$s = \bar{x} - x$$

and

$$y = \nabla_x L(\bar{x}, \bar{\lambda}, C) - \nabla_x L(x, \bar{\lambda}, C).$$

Assume that $x^* \in R^n$ is a solution of problem (1), which is regular (see Definition I.20). Let $\lambda^* \in R^m$ be the Lagrange multipliers guaranteed by Theorem I.8. It can be shown that under very mild conditions there exists $\hat{C} \geq 0$, such that the Hessian matrix $\nabla_x L(x^*, \lambda^*, C)$ is positive definite for any $C \geq \hat{C}$. If we assume that $C \geq \hat{C}$, and that f and g_i in problem (1) have continuous third-order partial derivatives, then we can state the following theorem. Its proof and the above details can be found in [2].

Theorem 1. If (x, λ) is sufficiently near (x^*, λ^*) and H is sufficiently near $\nabla_x^2 L(x^*, \lambda^*, C)^{-1}$, then the diagonalized Newton multiplier method and the diagonalized BFGS and DFP secant multiplier methods are well defined and each generates a sequence (x^k, λ^k) which converges to (x^*, λ^*). Moreover, for the diagonalized Newton multiplier method the convergence is quadratic in the sense that there exists M, such that

$$(16) \qquad \|(x^{k+1}, \lambda^{k+1}) - (x^*, \lambda^*)\| \leq M \|(x^k, \lambda^k) - (x^*, \lambda^*)\|^2, \qquad \forall k,$$

while for the two secant methods the convergence is superlinear in the sense that

$$(17) \qquad \lim_{k \to \infty} \frac{\|(x^{k+1}, \lambda^{k+1}) - (x^*, \lambda^*)\|}{\|(x^k, \lambda^k) - (x^*, \lambda^*)\|} = 0.$$

Consider problem (1) with $n = 3$, $m = 1$

$$f(x) = (x_1 - 1)^2 + (x_1 - x_2)^2 + (x_2 - x_3)^4$$

and

$$g(x) = x_1(1 + (x_2)^2) + (x_3)^4 - 4 - 3\sqrt{2}.$$

This problem has a relative minimizer at

$$\lambda^* = (1.1049, 1.1967, 1.5353),$$

with associated Lagrange multiplier $\lambda^* = -.0107$. Initialize as follows:

$$
\begin{aligned}
(18) \qquad x^0 &= (1.5, 1.5, 1.5) \\
\lambda^0 &= [\nabla g(x^0)^T \nabla g(x^0))^{-1} [g(x^0) - \nabla g(x^0)^t \nabla f(x^0)] \\
C^0 &= 0 \\
H^0 &= \nabla_x^2 L(x^0, \lambda^0, 0)^{-1}.
\end{aligned}
$$

Table II.1. Values of $\|x^k - x^*\|$

k	Newton	BFGS
1	$.31 \times 10^{-0}$	$.31 \times 10^{-0}$
2	$.14 \times 10^{-0}$	$.22 \times 10^{-0}$
3	$.35 \times 10^{-1}$	$.12 \times 10^{-0}$
4	$.29 \times 10^{-2}$	$.38 \times 10^{-1}$
5	$.19 \times 10^{-4}$	$.11 \times 10^{-1}$
6	$.90 \times 10^{-9}$	$.88 \times 10^{-3}$
7		$.29 \times 10^{-4}$
8		$.19 \times 10^{-5}$
9		$.44 \times 10^{-8}$

Note: The results of Table II.1 clearly indicate the quadratic and superlinear convergence phenomena.

Table II.1 compares the diagonalized Newton multiplier method with the diagonalized BFGS secant multiplier method for the initialization values given by (18).

II. 2. Optimization Problems with Nonnegativity Constraints

Let us consider problem (1) with addition of nonnegativity constraints. Specifically, we are interested in the constrained optimization problem

(19)   minimize $f(x)$;   subject to

$$g_i(x) = 0, \qquad i = 1, \ldots, m$$

and

$$x_i \geq 0, \qquad i = 1, \ldots, n,$$

where $f, g_i : R^n \to R$. Define the functions $\hat{f}, \hat{g}_i : R^n \to R$ by

(20) $\hat{f}(x_1, \ldots, x_n) = f(x_1^2, \ldots, x_n^2)$

and

(21) $\hat{g}_i(x_1, \ldots, x_n) = g_i(x_1^2, \ldots, x_n^2),$

and consider the equality constrained optimization problem

(22)   minimize $\hat{f}(x)$;   subject to $\hat{g}_i(x) = 0, \qquad i = 1, \ldots, m.$

Proposition 1.   If $x = (x_1, \ldots, x_n)$ solves problem (19), then $\sqrt{x} = (\sqrt{x_1}, \ldots, \sqrt{x_n})$ solves problem (22). Conversely, if $x = (x_1, \ldots, x_n)$ solves problem (22), then $x_1^2 = (x_1^2, \ldots, x_n^2)$ solves problem (19).

Proof. The proof follows almost by definition. ∎

Remark. Many authors feel that solving problem (19) by first solving problem (22) is undesirable and not the optimal approach. In principle, we certainly agree; however, we maintain that whether or not this approach should be used depends on the algorithm employed, as well as the particular problem. Moreover, we have found this approach to be very satisfactory for the problems described in Chapter 5 when using the diagonalized multiplier methods of Section II.1.

References

[1] Dennis, J. E., Jr., and Moré, J. J. (1977). "Quasi-Newton methods: motivation and theory." *SIAM Review* 10: 46–89.

[2] Tapia, R. A. (1977). "Diagonalized multiplier methods and quasi-Newton methods for constrained optimization." *Journal of Optimization theory and Applications* 22: 135–94.

Index

Akilov, G., 146

Bayes
  Theorem, 14
  Axiom, 15
  risk, 68
Bergstrom, H., 34
Biweight function, 36, 75
Bochner, S., 14, 55
Boneva, L., 85
Borel–Cantelli lemma, 127, 129
Brunk, H., 41, 42, 43

Carmichael, J., 84
Central Limit Theorem, 24, 29, 30
  proof of, 24–29
Closed set, 151
  weakly, 151
Compact set, 151
  weakly, 151
Consistency, 18
  of maximum likelihood estimator, 17
  of histogram, 46–48
  of kernel estimator, 55–57
  of discrete maximum penalized estimator, 124–29
Continuity, 149
  weak, 151
  semi-, 161
  weak semi-, 161
Contraction operator, 25
Convergence, 147

rate of kernel estimator, 59, 76
simulated rate of discrete maximum penalized likelihood estimator, 140
  weak, 151
  quadratic, 169
  superlinear, 169
  of diagonalized multiplier method, 169
Convex
  set, 151
  function, 154
    strictly, 154
    uniformly, 154
    differential characterizations of, 154–60
Cramér, H., 21

Daniel, J., 146
Davis, K., 78, 79
de Figueiredo, R., 71
De Moivre, 4
de Montricher, G., 102, 105
Dennis, J., 167
Density estimator
  maximum likelihood, 13, 15, 83, 92–101
  series, 36–44
  kernel, 44–82, 131–40
  histogram, 44–48
  shifted histogram, 50, 53, 54, 76
  modified histogram, 83
  histospline, 85
  generalized histogram, 96–99

Density estimator (*Continued*)
  maximum penalized likelihood, 102, 106, 113, 114–20
  de Montricher-Tapia-Thompson, 106–08
  Good-Gaskins first, 108–14
  Good-Gaskins second, 114–20
  Good-Gaskins pseudo, 115–19
  discrete maximum penalized likelihood, 121–43
  stable histogram, 124
  for multivariate densities, 141–43
Density function (distribution)
  normal (Gaussian), 4, 5, 10, 11, 12, 24, 29, 30, 35, 60
  hypergeometric, 5
  binomial, 5, 8
  Pearson family of, 5–11
  beta, 7, 8
  F, 9, 67, 79–82
  gamma, 9
  student's *t*, 11, 12
  multimodal, 12, 60–68, 131–37, 141, 143
  prior, 14, 15, 48
  posterior, 14, 15, 49
  uniform, 15
  Johnson family of, 30–33
  unimodal, 31, 34, 35, 37
  Cauchy, 33, 35
  symmetric stable, 33–36
  Dirac delta, 40, 41, 99, 105, 108, 111, 113
  Dirichlet, 49
  double exponential, 58
Diagonalized multiplier method, 166–71
Distributions (Schwartz), 105, 111, 113, 148
Dual space, 149
Dunford, H., 146

Efficiency, 21, 35, 67, 75, 77, 78
Elderton, W., 5, 7
Epanechnikov, V., 73, 76

Fama, E., 33, 34
Farrell, R., 71
Feller, W., 33
Fisher, R., 5, 13, 14, 18, 21, 22, 23
Fourier transform, 57, 69, 77, 79, 120
Functional iteration, 43

Galton, F., 24, 29, 30
Gaskins, R., 86, 87, 88, 89, 104, 105, 108, 109, 110, 113, 114, 115, 117, 119, 120

Gâteaux differentiation, 103, 118, 154–55
Goffman, C., 146
Goldstein, A., 146
Good, I., 86, 87, 88, 89, 104, 105, 108, 109, 110, 113, 114, 115, 117, 119, 120
Gosset, W., 11, 12
Gradient, 107, 111, 113
  definition of, 155
Gram-Charlier Representation, 37, 38
Graunt, J., 3, 4

Halley, 4
Henry VIII, 2
Hermite polynomials, 37, 87
Hilbert space
  definition of, 147
  reproducing kernel, 152
  proper functional, 152
Histogram, 2, 4, 5, 44
  as maximum likelihood estimator, 45, 95–96
  consistency, 46–48
  shifted, 50–54
  modified, 83
  generalized, 98
  stable, 124
Horvath, J., 105
Huygens, 4

IMSL, 130
Inequality
  Cauchy-Schwarz, 9, 19, 41, 147
  Jensen, 16
  Cramer-Rao, 19, 20, 21
  Chebyshev, 87, 127, 128
  triangle, 147
Infinity property, 162
Inner product space, 123, 148
  definition, 146
  properties, 146

Johnson, N., 5, 7, 13, 31, 32, 33

Kantorovich, L., 146
Kazakos, D., 73
Kendall, M., 5, 18, 85
Kernel, 75
  density estimator, 44
  characteristic exponent, 58, 67, 75, 78, 84
  characteristic coefficient, 58, 78
  examples of, 60
  Gaussian, 60, 75, 76, 138–40

quasi-optimal support, 67, 68
scaling of support, 131–40
Kronmal, R., 38, 39
Kuhn-Tucker conditions, 164

Lagrange multipliers, 88, 107, 113, 119, 125
    necessity conditions, 163–64
    update formula, 168
Lagrangian, 130
    augmented, 167
LARYS, 143
Leadbetter, M., 76, 77
Lii, K., 86
Likelihood, 14, 92
    maximum, 13, 20, 21, 22, 93, 95
    maximum penalized, 102, 121
    penalized, 102–03
    discrete maximum penalized, 124
Lindeberg, W., 24, 29, 30
Lions, J., 148
Luenberger, D., 146, 164

Magenes, E., 148
Mean Square Error, 46, 47, 48, 50, 52, 53, 54, 56, 57, 59, 72, 73, 74
    integrated, 40, 41, 48, 54, 59, 67, 75, 76, 77, 78, 138, 139, 140
Method of moments, 7, 13, 21, 38
Monte Carlo, 34, 67–68, 129–40
Moré, J., 167
Moses, 1, 2

NASA, 143
Newton's method, 67, 130, 168
    quasi, 167–70

Oden, J., 148
Operator, 150
    definition of, 147
    linear, 147
    bounded, 149
    continuity of, 149
    norm, 149
Optimization problem
    examples of, 45, 76, 77, 86, 88, 93, 94, 102, 104, 107, 109, 110, 114, 115, 118, 123, 124
    existence of solutions to, 160–63
    uniqueness of solutions to, 160–63
    characterizations of solution to, 164
    numerical solution, 166–71
    with nonnegativity constraints, 170–71
Ord, J., 5
Ortega, J., 154

Parzen, E., 54, 55, 59, 71, 84
Pascal, 3
Pearson, K., 4, 5, 13, 18, 21, 22, 32
Pedrick, G., 146
Penalty
    function, 88, 102–20, 130
    constant, 167–71
Petty, 4
Princeton Robustness Study, 35

Rall, L., 146
Reddy, J., 148
Regular constraints, 164
Rényi, A., 24
Remote sensing, 142–45
Reproducing kernel Hilbert space, 103, 106, 116, 152–54
Rheinboldt, W., 154
Riesz
    representer, 106, 107
    representation theorem, 150
Roll, R., 33, 34
Rosenblatt, M., 50, 53, 54, 57, 73, 76, 86
Royden, M., 146

Schoenberg, I., 85, 86, 108
Schwartz, J., 146
Scott, David, 75, 121, 124, 131
Second derivative, 154
    positive definite, 157
    positive semidefinite, 157
    uniformly positive definite, 157
Shapiro, J., 78
Shenton, L., 38
Sobolev space, 105, 106, 108, 109, 110, 111, 113, 114, 116, 118, 120
    discrete, 123
    definition, 148
Splines
    B-splines, 44, 76
    histospline, 85
    splinegram, 86
    polynomial, 98, 106, 108, 122
    monospline, 106, 108, 124
    exponential, 111, 113, 114
Stability, 33
    dimensional, 100
Stationarity, 69, 70, 71, 84
Stefanov, I., 85
Strong Law of Large Numbers, 17, 129
Stuart, A., 5, 18

Tangent cone, 157
Tapia, R., 102, 105, 121, 124, 130, 131, 154, 166
Tarter, M., 38, 39

Taylor, A., 146, 164
Thompson, J., 102, 105, 121, 124, 131
Tukey, J., 35, 75, 79

van Dael, 4
Van Ryzin, J., 53

Wahba, G., 71, 73, 74, 76, 131

Wald, A., 18
Watson G., 41, 76, 77
Wegman, E., 83
Whittle, P., 68, 69, 70
Wilks, S., 16

Yule–Walker equations, 84

THE JOHNS HOPKINS UNIVERSITY PRESS

This book was composed in Monophoto Times Roman text and display type by Syntax International from a design by Susan Bishop. It was printed on 50-lb. Publishers Eggshell Wove paper and bound in Joanna Arrestox cloth by Universal Lithographers, Inc.

Library of Congress Cataloging in Publication Data

Tapia, Richard A.
  Nonparametric probability density estimation.

  (Johns Hopkins series in the mathematical sciences; no. 1)
  Includes index.
  1. Distribution (Probability theory) 2. Estimation theory. 3. Nonparametric statistics. I. Thompson, James R., 1938–   joint author. II. Title. III. Series.

QA273.6.T37   519.2′4   77–17249
ISBN 0–8018–2031–6

AMS